THE PHYSIOLOGY AND BIOCHEMISTRY OF PLANT RESPIRATION

Edited by

J. M. PALMER

Reader in Enzymology
Imperial College of Science and Technology
University of London

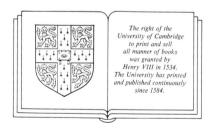

The right of the
University of Cambridge
to print and sell
all manner of books
was granted by
Henry VIII in 1534.
The University has printed
and published continuously
since 1584.

CAMBRIDGE UNIVERSITY PRESS

Cambridge

London New York New Rochelle

Melbourne Sydney

Published by the Press Syndicate of the University of Cambridge
The Pitt Building, Trumpington Street, Cambridge CB2 1RP
32 East 57th Street, New York, NY 10022, USA
296 Beaconsfield Parade, Middle Park, Melbourne 3206, Australia

© Cambridge University Press 1984

First published 1984

Printed in Great Britain by the University Press, Cambridge

Library of Congress catalogue card number: 83-26279

British Library Cataloguing in Publication Data

The Physiology and biochemistry of plant respiration – (Society for Experimental
Biology seminar series, 20)
1. Plants – Respiration
I. Palmer, J. M.   II. Series
581.1′2   QK891

ISBN 0 521 23697 5

CONTENTS

CONTRIBUTORS

ap Rees, T.
Botany School, University of Cambridge, Cambridge CB2 3EA, UK

Bergman, A.
Department of Biochemistry, University of Umeå, S-901 87 Umeå, Sweden

Bervillé, A.
Station d'Amélioration des Plantes, INRA-BV 1540, 21034 Dijon Cedex, France

Cooke, A.
Department of Botany, Manchester University, Manchester M13 9PL, UK

Crawford, R. M. M.
Department of Botany, St Andrews University, St Andrews, Fife KY16 9AL, UK

Davies, D. D.
School of Biological Sciences, University of East Anglia, Norwich NR4 7TJ, UK

Earnshaw, M. J.
Department of Botany, Manchester University, Manchester M13 9PL, UK

Ericson, I.
Department of Biochemistry, University of Umeå, S-901 87 Umeå, Sweden

Gardeström, P.
Department of Biochemistry, University of Umeå, S-901 87 Umeå, Sweden

Frenkel, C.
Department of Horticulture and Forestry, Cook College, Rutgers State University, PO Box 231, New Brunswick, New Jersey 08903, USA

Gauvrit, C.
Laboratoire de Malherbiologie, INRA-BV 1450, 21034 Dijon Cedex, France

Meeuse, B. J. D.
Room 408, Johnson Hall, Department of Botany, AJ-10, University of Washington, Seattle, Washington 98195, USA

Møller, I. M.
Department of Plant Physiology, University of Lund, Box 7007, S-220 Lund, Sweden

Moore, A. L.
Department of Biological Sciences, University of Sussex, Falmer, Brighton BN1 9QQ, UK

Paillard, M.
Station d'Amélioration des Plantes, INRA-BV 1540, 21034 Dijon Cedex, France

Palmer, J. M.
Department of Pure and Applied Biology, Imperial College, Prince Consort Road, London SW7 2BB, UK

Pradet, A.
Station de Physiologie Végétale, INRA Centre de Recherches de Bordeaux, 33140 Pont de la Maye, France

Raymond, P.
Station de Physiologie Végétale, INRA Centre de Recherches de Bordeaux, 33140 Pont de la Maye, France

Ryle, G. J. A.
ARC Grassland Research Institute, Hurley, Maidenhead SL6 5LR, UK

Simon, E. W.
Department of Botany, Queen's University, Belfast BT7 1NN, UK

van der Plas, L. H. W.
Biologisch Laboratorium, Vrije Universiteit, de Boelelaan 1087, Amsterdam 1007mc, The Netherlands

Wilson, S. B.
Department of Biochemistry, Marischal College, University of Aberdeen, Aberdeen AB9 1AS, UK

PREFACE

The behaviour of cellular respiration during the physiological development of plants and their response to environmental stress is an important topic in plant physiology. In recent years research has concentrated on the detailed biochemistry of isolated mitochondria and several specialist books have been published on this aspect of the subject. A central objective when organising the symposium upon which the chapters in this book are based was to combine topics in the physiology and biochemistry of plant respiration. When preparing their manuscripts the authors were asked to present them to an advanced undergraduate readership rather than to established research workers.

In both Part I on physiology and Part II on biochemistry, an attempt has been made to present some general topics as mini-reviews and to include some more specialised topics as shorter chapters. Although the meeting associated with this publication was held at the end of 1980 most of the papers have been updated to early 1983.

July 1984   J.M.P.

PART I

The physiology of plant respiration

G. J. A. RYLE

1 Respiration and plant growth

Respiration – most familiar through the sequence of reactions whereby sugars and other substrates are oxidised to carbon dioxide and water – generates two different kinds of products essential to the integrity and growth of the plant. The first of these is energy released in a form which can be utilised to synthesise the major groups of compounds (proteins, carbohydrates, fats, etc.) which together constitute the tissues of the plant. Such energy is also available for all aspects of metabolic transport and accumulation. The energy stored in the bonding of the original substrate molecule is released in a controlled manner by the step-by-step breakdown of the molecule through sequential enzyme action. Reduced nucleotides and ATP, which can be constantly regenerated, are the agents whereby the energy is temporarily trapped before use in specific reactions.

The second product of respiration is the array of intermediates or carbon skeletons produced during the stepwise breakdown of the substrate molecules. These are the basic building units from which new plant tissues are synthesised. Assimilate entering respiratory pathways contributes to both products. A corollary is that not all substrate molecules are oxidised to water and carbon dioxide as summarised in the classic equation:

$$C_6H_{12}O_6 + 6O_2 \rightarrow 6CO_2 + 6H_2O.$$

The analysis and measurement of respiration in whole plants

In whole plants, and still more in communities of plants, little or no detailed information can be obtained on the flux of energy through respiratory pathways, and an understanding of the nature and magnitude of respiration has been sought in measurements of loss of carbon dioxide, losses which are often masked by contrary fluxes of carbon dioxide into photosynthetic pathways. It is not perhaps surprising, therefore, that the study of respiration in relation to growth and yield in agriculture lagged behind that of photosynthesis. A more balanced approach is now apparent, largely as a result of a series of related advances in knowledge which coalesced in the late 1960s.

Earlier, attempts to reconcile photosynthesis and growth in plant systems encountered difficulties which ultimately stemmed from an inherent assumption that respiratory losses of carbon dioxide were closely related to plant biomass, expressed variously as leaf area, leaf area index or dry weight. In a series of careful investigations in controlled environments, the rates of respiration of leaves in canopies of a range of crop plants were measured in a variety of light and temperature regimes (Ludwig, Saeki & Evans, 1965; McCree & Troughton, 1966a, b). Rate of respiration was found to vary with environment and to adapt quickly to changes in specific environmental factors. Furthermore, McCree & Troughton (1966b) went on to postulate that respiration in such artificial plant communities was closely linked to the current rate of photosynthesis, and that there was probably a second component of respiration related, as before, to plant biomass. Subsequently, these concepts were refined and expressed in a simple mathematical formula which still provides the conceptual basis for the contemporary view of respiration in plants and crops (McCree, 1970). McCree expressed the dual aspects of respiration in the relationship:

$$R = kP + cW,$$

where R is the 24 h total of respiration (g CO_2 d^{-1} per plant), k is a dimensionless constant, P is daily gross photosynthesis (g CO_2 d^{-1} per plant), c is a constant with the dimension time^{-1}, and W is the dry weight of the live plant material (g CO_2 per plant). The idealised plots in Fig. 1(a) and (b) illustrate the basic relationships. In McCree's experiments, where the plants were grown in constant environments except for the consecutive daily changes in light intensity, the value for the constant k was found to be 0.25 and that for c to be 0.015 per day. In other words, the total respiratory losses of carbon dioxide in one day were proportional to 25% of the day's photosynthate plus a further respiratory loss equivalent to 1.5% of the dry weight of live plant material (biomass) present on the same day.

These concepts of a respiratory loss linked with the synthesis of new tissues and another related to the activities of mature tissue, were utilised elsewhere. Hesketh and co-workers (Hesketh, Baker & Duncan, 1971) derived similar relationships while developing a quantitative model for the cotton crop, while related concepts had been used earlier by Hiroi & Monsi (1964) and de Wit, Brouwer & Penning de Vries (1970). Analyses of the separate functions of respiration in animal physiology and in bacterial growth provided other sources for very similar conclusions (Thornley, 1970). Indeed, farmers have long operated animal feeding systems based on starch equivalent or, more recently, metabolisable energy inputs which make a clear distinction between the energy required for unit live-weight gain or litre of milk production and for the maintenance of the animal.

The emphasis upon a component of respiration linked to the use of photosynthate in growth and another related in some way to existing biomass had a biochemical counterpart in the analysis of respiratory functions. Beevers (1970) considered how respiration participates in synthetic events (the growth of new tissues), in protein turnover and the maintenance of tissue integrity (maintenance), and how respiration may occur without any useful product being generated for the plant (uncoupled or idling respiration). However, it was Penning de Vries (1972, 1975a, b; Penning de Vries, Brunsting & van Laar, 1974) who provided the quantitative biochemical framework for these respiratory relationships when they published their analyses of the fluxes of carbon through known biochemical pathways, and the concurrent energy costs, during the synthesis of new plant tissues and the maintenance of existing ones. Their method consisted of using the most efficient known biochemical pathways to calculate the weight of each class of plant product which could be synthesised from unit weight of substrate, taking account both of material changes and of the chemical energy required. If the nature of the energy substrate and the plant's growth products were known, the numerical relationship between weight of substrate, weight of

Fig. 1. (*a*) Generalised relationship between gross photosynthesis and the 24 h total of respiration in four plants of white clover of increasing weight numbered 1–4. Each plant was cycled through single days of increasing or decreasing light intensity to determine rates of photosynthesis and respiration characteristic of each light regime. (*b*) Generalised relationship between live plant weight and the 24 h total of respiration at zero photosynthesis in white clover, obtained by plotting the intercept values of (*a*) against plant dry weight. (After McCree, 1970.)

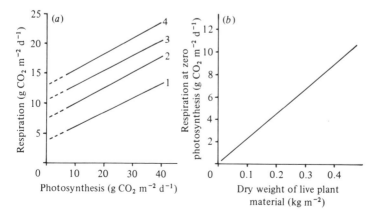

product, carbon dioxide respired, oxygen consumed and chemical energy required could all be calculated (Table 1). In the same way, the material and energy costs of maintaining tissue integrity in mature plant tissues could also be estimated. Good correspondence between estimate and experiment was obtained in relation to synthetic processes and, moreover, the conversion efficiency calculated for the total of synthetic reactions in several plants agreed closely with that measured in white clover by McCree (1970).

The respiratory costs associated with energy use in mature tissues proved more difficult to estimate due to the problems of defining the range and scale of molecular turnover and other costs in tissues in this state. Very similar problems have been encountered in relation to defining the scope of basal metabolism, or maintenance, in animal physiology (Brody, 1945). Even so, the estimates and measurements of maintenance respiration, although more variable, were also similar to those reported by McCree (1970).

It is not possible using infra-red gas analysis to distinguish directly the carbon dioxide efflux from the biosynthesis of new tissue from that generated by mature tissues, because they occur simultaneously during normal growth. This problem can be partially overcome by using radiocarbon to label the assimilate formed during photosynthesis. Under favourable growing conditions most of the labelled assimilate is quickly entrained in the biosynthesis of new tissues and the concomitant efflux of $^{14}CO_2$ can be determined. Subsequently, the new tissues synthesised from the labelled assimilate mature and their labelled respiratory efflux can then also be measured. As with other techniques, no absolute distinction can be drawn because one process merges imperceptibly into the other. Further, some underestimate of respiration is

Table 1. *The amount of glucose required to synthesise 1 g of plant tissue in 25-day-old maize plants*

Compound or process	Weight of compounds (g)	Glucose required (g)	Carbon dioxide produced (g)	Oxygen consumed (g)
Organic N compounds	0.23	0.478	0.272	0.0148
Lipids	0.025	0.070	0.033	0.0025
Carbohydrates	0.64	0.735	0.047	0.0345
Carboxylic acids	0.04	0.028	−0.007	0.0036
Mineral uptake	0.065	0.003	0.005	0.0035
Glucose uptake	—	0.070	0.103	0.0750
Total	1.00	1.384	0.453	0.1339

From Penning de Vries (1972).

likely because of refixation of $^{14}CO_2$ in the illuminated shoot. Nevertheless, this technique provides a means of separating the two processes in time.

Tanaka and co-workers (Tanaka, Kawano & Yamaguchi, 1966) showed in field-grown rice that 35–50% of the carbon in labelled assimilate was respired during the life of the crop, most of it in the first few days after its formation in the leaf. Other more recent investigations with a range of graminaceous species both temperate and tropical have shown that where labelled assimilate is used for the synthesis of leaf, stem and root, 25–35% of the carbon is quickly lost in respiration (Ryle, Brockington, Powell & Cross, 1973; Ryle & Powell, 1974). Most of this initial burst of respiration provides the energy to mobilise the labelled assimilate and transport it to meristems, and to provide carbon skeletons and chemical energy to synthesise

Fig. 2. Generalised relationship between respiration and the utilisation of ^{14}C-labelled assimilate during vegetative growth in a graminaceous plant. Data are for a young fully expanded leaf exposed to $^{14}CO_2$ during normal photosynthesis for 30 min.

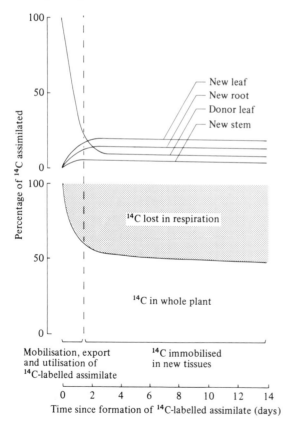

new tissues (Gordon, Ryle & Powell, 1977). Direct measurements show that only a small proportion of this initial $^{14}CO_2$ loss is associated with events in the photosynthesising leaf and that most is the result of synthetic processes (growth) in the meristems. These processes are largely complete in 1–2 days in favourable growing conditions (Fig. 2). Subsequently, respiratory loss of the labelled carbon atoms (as $^{14}CO_2$) continues, as the tissues formed from the meristems mature and senesce, but at a very slow rate compared with that characteristic of the first 1–2 days. Circumstantial evidence indicates that such respiratory losses are associated with protein turnover and the maintenance of unstable cell structures (Beevers, 1970). These losses of labelled carbon from mature and ageing tissues amount to a further *c.* 20%, so that in total *c.* 50% of the carbon in the original assimilate molecules is lost in respiration. The loss from ageing tissues can also be expressed on a daily basis per unit of tissue: the values fall to a minimum of 10–20 mg CH_2O g^{-1} d^{-1} after 7–14 days.

Respiratory effluxes in plants and crops

Since its formulation in 1970, the two-component concept of respiration has been modified and adapted to deal with a wide range of experimental situations, while efforts to refine its theoretical basis have continued (Thornley, 1977; Barnes & Hole, 1978).

As far as that component of respiration assigned to synthetic processes is concerned, most investigators agree that neither light nor temperature within the range normally encountered markedly influences the calculated efficiency of biosynthesis: in both temperate and tropical plants efficiency generally falls within the range 65–80% (McCree, 1970, 1974; Penning de Vries *et al.*, 1974; Ryle, Cobby & Powell, 1976). However, it has been pointed out that root growth has a lower conversion efficiency than shoot growth; and further that assumptions concerning the roles of photophosphorylation and phosphorylation in the illuminated shoot, and the tightness of coupling of growth and photosynthesis, will influence the calculated efficiency of biosynthesis (Penning de Vries, 1972; Hansen & Jensen, 1977). Lack of mineral nitrogen or the development of reproductive organs can also influence conversion efficiency; in these situations, temporary or permanent alterations in the nature of the growth products have been invoked (Yamaguchi, 1978).

The component of respiration assigned to biomass maintenance appears to be sensitive to a range of factors. Experimental values for a variety of whole plants including beans, soyabean, grasses, clovers, sunflower, maize and sorghum in normal growing conditions vary between 5 and 37 mg CH_2 g^{-1} dry wt d^{-1} (McCree, 1970, 1974; Penning de Vries, 1972; Ryle *et al.*, 1976; Hansen & Jensen, 1977; Yamaguchi, 1978; Jones, Leafe, Stiles & Collett,

1978). The response to temperature in the short-term exhibits a Q_{10} of 1.8–2.2 (McCree, 1974; Penning de Vries, 1972, 1975a, b; Ryle *et al.*, 1976). The long-term interrelationships between enhanced respiration rates and the acceleration of tissue ageing and senescence, over the life of a leaf or root for example, are poorly documented (Sheehy, Cobby & Ryle, 1980). Measured rates of maintenance respiration appear also to vary with level of mineral nutrition, plant species, rate of plant growth and stage of plant development (McCree, 1974; Penning de Vries, 1975a, b; Ryle *et al.*, 1976; Yamaguchi, 1978; Jones *et al.*, 1978).

That the rates of maintenance respiration vary with plant and environment is not perhaps surprising, since the term 'maintenance' covers a broad range of biochemical processes and metabolic functions, the rates of which are ill defined in whole plants. Moreover, as emphasised earlier, the experimental techniques most used for estimating the respiration associated with the biosynthesis of new tissues and the maintenance of mature tissues can only provide operational distinctions which are by no means definitive. The real value of such distinctions lies in their power to provide an understanding of how the primary process of photosynthesis feeds through to growth and yield.

Rates of maintenance respiration are generally expressed as costs per unit plant biomass per unit time because they are commonly sought as simple terms which can be inserted in an appropriate mathematical expression to predict plant or crop growth. In young plants, where the accumulated biomass of mature tissue is small, the product of maintenance rate and weight of biomass is also small relative to the photosynthetic flux into the plant. Most experimental observations indicate that during much of their early growth plants exhibit a total respiratory efflux of about 40–50 % of their estimated gross photosynthesis, of which half or less is generally attributed to maintenance processes (see literature already cited). As plant biomass increases, particularly in crop environments where light interception reaches an optimum and interplant competition is intense, the proportion of assimilate respired for maintenance might also be expected to continue to increase. Some increase does occur, but several factors limit its extent. Plant tissues such as leaves and roots die and are shed, while those having a predominantly structural function such as stem respire only slowly after having attained maturity. For example, in a grass monoculture growing vegetatively, a dynamic equilibrium appears to be attained in which rate of photosynthesis, rate of new growth, maintenance and growth respiration, and plant biomass are all constant (Fig. 3). New growth is balanced by the death of old tissues, and the maintenance respiration load is constant because the average age of plant biomass, and its composition, is also constant. Even in such a 'conceptually' simple crop as this, the equilibrium situation depicted in Fig. 3 can only be

Fig. 3. Generalised relationship between photosynthesis and respiration in an S24 perennial ryegrass sward during vegetative growth in a controlled environment.

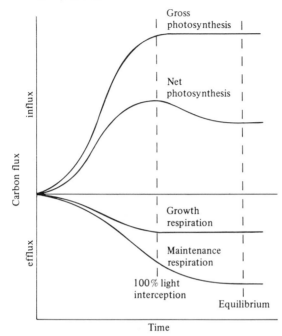

Fig. 4. Changes in maintenance respiration and dry weight during the growth of a reproductive S24 perennial ryegrass crop. Maintenance respiration was calculated on a dry weight basis (■) and a protein basis (□). ●, dry weight. (After Jones *et al.*, 1978.)

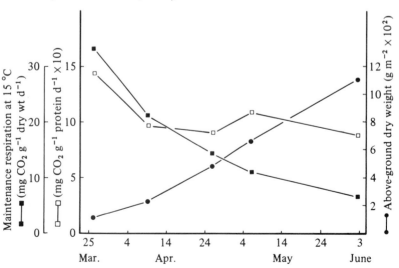

postulated for a constant environment. In the field, the constantly shifting environment, and morphological and physiological changes within the crop, prevent any lasting equilibrium being achieved.

In the more complex but perhaps more normal crop situation, where the grass develops elongated stems and inflorescences during early summer, the average maintenance respiration rate of the above-ground biomass appears to decrease progressively from $c.$ 30 to 7 mg CO_2 g^{-1} dry wt d^{-1} as the crop matures (Jones *et al.*, 1978). If the calculation base is shifted to unit protein in the biomass, to take some account of the accumulation of mature structural material, a more uniform rate of maintenance respiration is regained (Fig. 4).

During the later stages of development of most grain and grain legume crops the relationships between photosynthesis, respiration and growth also become very complex as the harvestable yield is progressively concentrated in the seeds and seed heads, photosynthesis declines and the distinction between growth and maintenance respiration becomes increasingly blurred. What does emerge is that at maturity the total respiratory turnover in crops such as maize, rice, soyabean and barley approaches 80–90% of the photosynthetic uptake of carbon (Biscoe, Scott & Monteith, 1975; Yamaguchi, 1978). The energy costs of synthesising unit weight of the harvest products of a variety of crop plants have been calculated on the basis of the biochemical steps most probably involved (Penning de Vries, 1972; Yamaguchi, 1978); in the most extreme comparison they differ by a factor of about 2 (Table 2). However, it is still not clear how far such differences in biosynthetic energy costs contribute to the total respiratory burden at maturity (compared with

Table 2. *Relative yielding ability of various crops estimated from chemical composition and 'conversion efficiency'*

	Chemical content (%)				Yielding ability (g per 100 g glucose)
Crop	Protein	Lipid	Carbo-hydrate	Ash	
Rice	8.8	2.7	87.0	1.5	73
Maize	9.5	5.3	83.7	1.5	70
Wheat	12.1	2.3	83.7	1.9	71
Barley	11.6	2.2	83.4	2.8	71
Soyabean	39.0	19.9	35.4	5.7	45
Field bean	24.1	2.6	69.0	4.3	62
Potato	9.3	0.5	86.3	3.9	75
Sesame	21.2	54.7	18.4	5.7	37

From Yamaguchi (1978).

normal biomass respiration), the costs of mobilising and utilising stored substances and other plant constituents and, possibly, an increasing component of respiration not coupled to any useful function in the plant (Beevers, 1970; Yamaguchi, 1978).

Respiratory efficiency in plants and crops

As the physiological understanding of respiration in whole-plant systems has improved, respiratory burdens have been increasingly scrutinised in the search for potentially more efficient crop plants. In the light of present biochemical knowledge, there appears to be little scope for improving the efficiency of biosynthetic processes. Much less is known about the energy costs of maintaining metabolic integrity in mature tissues. Differences in the rate of tissue respiration have been reported (Mooney & Billings, 1961), and in some instances growth rate has been shown to be inversely related to respiration rate (Heichel, 1971). More recently, Wilson (1975) has selected perennial ryegrass for fast and slow dark respiration (normal night) of mature leaf blades and established promising heritability for this trait. Since mature leaf blades of the grasses exhibit no growth functions, their night respiration rates presumably reflect maintenance processes plus the costs of mobilising and loading assimilate accumulated during the previous photoperiod into translocatory pathways. Experimental analysis of the growth of this plant material showed that 'low' respiration selections outyield (6–13%) both 'high' respiration and parent populations in the growth room (Robson, 1982) and in field trials (Wilson & Jones, 1982).

Another recent example of the scrutiny brought to bear on the respiratory burdens of plants concerns the energy costs which arise when legumes reduce and metabolise the dinitrogen molecule in root nodules. Biochemical studies indicate a large energy requirement in the form of ATP and reductants for the synthesis in the root nodule of ammonia from gaseous nitrogen (Bulen & LeComte, 1966), while very similar costs are also thought to be incurred when nitrate – the most common form of combined nitrogen in the soil – is reduced in the plant to ammonia (Bergersen, 1971). Experimental evaluation of these concepts in whole plants by comparing the respiratory burdens of plants fixing their own nitrogen in root nodules with those of plants utilising nitrate taken up by roots, has revealed a respiratory cost of the appropriate magnitude which can be assigned to nitrogen fixation in the root nodule, but no clear evidence of a corresponding respiratory efflux for nitrate reduction has been adduced in plants utilising nitrate-nitrogen (Mahon, 1977; Silsbury, 1977; Ryle, Powell & Gordon, 1979).

A role for light in nitrate reduction has long been accepted and both ATP and reductants generated in photosynthesis may participate in some of the reactions involved if the site of reduction is the illuminated leaf (Heber, 1974; Miflin & Lea, 1976). The implications in relation to the energy costs of synthesising new tissue were considered by Penning de Vries (1975a). The lack of clear evidence concerning the intensity of nitrate reduction in the roots and shoots of crop plants presently hinders evaluation of the energy costs in terms either of competition with carbon dioxide reduction in photosynthesis or of the utilisation of assimilate already formed. The existing experimental evidence indicates that in bright light vigorously growing plants, provided with an optimum supply of nitrate-nitrogen, respire less and grow more than their counterparts fixing their own nitrogen (Schubert & Ryle, 1980).

Oxidative pathways of low phosphorylative capacity have been widely studied in isolated mitochondria (Solomos, 1977). In whole plants a role has been occasionally accorded them in senescing tissues or during periods of stress (Beevers, 1970; Yamaguchi, 1978), but in the main it has been implicitly assumed that the respiration of most plants and crops is generally coupled tightly to growth processes. This tacit assumption has been strengthened by much experimental data, already referred to, where respiration and growth in whole plants do appear to be quantitatively related in the manner predicted by 'most efficient biochemistry' calculations.

Recent investigations of the relationship between root growth and root respiration, using the uptake of oxygen as the measure of respiratory activity, indicate the involvement of an oxidase (salicyl hydroxamic acid SHAM-sensitive), other than cytochrome oxidase, which may account for 40–70% of the root respiration of *Senecio* species (Lambers & Smakman, 1978). In the shoot of *Senecio* the SHAM-sensitive oxidase appears to be inoperative (Lambers, Noord & Posthumus, 1979). Lambers and co-workers conclude that assimilate is wasted in the roots of *Senecio* and other species, partly by the SHAM-sensitive pathway, but also sometimes via a further pathway whose biochemical nature is unknown (Lambers & Steingrover, 1978).

That roots exhibit higher unit respiration rates than shoots is documented (Hansen & Jensen, 1977; Lambers *et al.*, 1979; Ryle *et al.*, 1979), although the metabolic basis for this is still not clear. Since plants often possess much less root than shoot, high unit respiration rates in roots could be obscured in *whole-plant* analyses without substantially altering the total calculated costs of synthesising unit amount of plant tissue. A further unknown factor is that of the participation of photophosphorylation in shoot growth (Heber, 1974; Raven, 1976). If this source of energy is substantial, analyses using whole-plant carbon dioxide effluxes will underestimate the true energy costs of shoot growth but, at the same time, perhaps numerically compensate for high (wasteful) respiration in the root.

The resolution of these problems requires more information concerning the flux of metabolites between root and shoot, the costs of absorbing and utilising macronutrients, and the relative roles of photophosphorylation and phosphorylation in shoot growth.

References

Barnes, A. & Hole, C. C. (1978). A theoretical basis of growth and maintenance respiration. *Annals of Botany*, **42**, 1217–21.

Beevers, H. (1970). Respiration in plants and its regulation. In *Prediction and Measurement of Photosynthetic Productivity*, ed. I. Malek, pp. 209–14. Wageningen: Centre for Agricultural Publishing and Documentation.

Bergersen, F. J. (1971). The central reactions of nitrogen fixation. *Plant and Soil (Special Volume)*, 511–24.

Biscoe, P. V., Scott, R. K. & Monteith, J. L. (1975). Barley and its environment. III. Carbon budget of the stand. *Journal of Applied Ecology*, **12**, 269–93.

Brody, S. (1945). *Bioenergetics and Growth*. New York: Reinhold.

Bulen, W. A. & LeComte, J. R. (1966). The nitrogenase system from azotobacter: two enzyme requirement for N_2 reduction, ATP-dependent H_2 evolution, and ATP hydrolysis. *Proceedings of the National Academy of Sciences, USA*, **56**, 979–86.

Gordon, A. J., Ryle, G. J. A. & Powell, C. E. (1977). The strategy of carbon utilization in uniculm barley. I. The chemical fate of photosynthetically assimilated ^{14}C. *Journal of Experimental Botany*, **28**, 1258–69.

Hansen, G. K. & Jensen, C. R. (1977). Growth and maintenance respiration in whole plants, tops, and roots of *Lolium multiflorum*. *Physiologia Plantarum*, **39**, 155–64.

Heber, U. (1974). Metabolite exchange between chloroplasts and cytoplasm. *Annual Review of Plant Physiology*, **25**, 393–421.

Heichel, G. H. (1971). Confirming measurements of respiration and photosynthesis with dry matter accumulation. *Photosynthetica*, **5**, 93–8.

Hesketh, J. D., Baker, D. N. & Duncan, W. G. (1971). Simulation of growth and yield in cotton: respiration and carbon balance. *Crop Science*, **11**, 394–7.

Hiroi, T. & Monsi, M. (1964). Physiological and ecological analyses of shade tolerance of plants. IV. Effects of shading on distribution of assimilate in *Helianthus annuus*. *Botanical Magazine, Tokyo*, **77**, 1–9.

Jones, M. B., Leafe, E. L., Stiles, W. & Collett, B. (1978). Pattern of respiration of a perennial ryegrass crop in the field. *Annals of Botany*, **42**, 693–703.

Lambers, H., Noord, R. & Posthumus, F. (1979). Respiration of *Senecio* shoots: inhibition during photosynthesis, resistance to cyanide and relation to growth and maintenance. *Physiologia Plantarum*, **45**, 351–6.

Lambers, H. & Smakman, G. (1978). Respiration of flood-tolerant and flood-intolerant *Senecio* species as affected by low oxygen tension. *Physiologia Plantarum*, **42**, 163–6.

Lambers, H. & Steingrover, E. (1978). Growth respiration of a flood-tolerant and a flood-intolerant *Senecio* species: correlation between calculated and experimental values. *Physiologia Plantarum*, **43**, 219–24.

Ludwig, L. J., Saeki, T. & Evans, L. T. (1965). Photosynthesis in artificial communities of cotton plants in relation to leaf area. I. Experiments with progressive defoliation of mature plants. *Australian Journal of Biological Sciences*, **18**, 1103–18.

McCree, K. J. (1970). An equation for the rate of respiration of white clover plants grown under controlled conditions. In *Prediction and Measurement of Photosynthetic Productivity*, ed. I. Malek, pp. 221–30. Wageningen: Centre for Agricultural Publishing and Documentation.

McCree, K. J. (1974). Equations for the rate of dark respiration of white clover and grain sorghum, as functions of dry weight, photosynthetic rate, and temperature. *Crop Science*, **14**, 509–14.

McCree, K. J. & Troughton, J. H. (1966*a*). Prediction of growth rate at different light levels from measured photosynthesis and respiration rates. *Plant Physiology*, **41**, 559–66.

McCree, K. J. & Troughton, J. H. (1966*b*). Non-existence of an optimum leaf area index for the production rate of white clover grown under constant conditions. *Plant Physiology*, **41**, 1615–22.

Mahon, J. D. (1977). Respiration and the energy requirement for nitrogen fixation in nodulated pea roots. *Plant Physiology*, **60**, 817–21.

Miflin, B. J. & Lea, P. J. (1976). The pathway of nitrogen assimilation in plants. *Phytochemistry*, **15**, 873–85.

Mooney, H. A. & Billings, W. D. (1961). Comparative physiological ecology of arctic and alpine populations of *Oxyria digynia*. *Ecological Monographs*, **31**, 1–28.

Penning de Vries, F. W. T. (1972). Respiration and growth. In *Crop Processes in Controlled Environments*, ed. A. R. Rees, K. E. Cockshull, D. W. Hand & R. G. Hurd, pp. 327–48. New York & London: Academic Press.

Penning de Vries, F. W. T. (1975*a*). Use of assimilates in higher plants. In *Photosynthesis and Productivity in Different Environments*, ed. J. P. Cooper, pp. 459–80. Cambridge University Press.

Penning de Vries, F. W. T. (1975*b*). The cost of maintenance processes in plant cells. *Annals of Botany*, **39**, 77–92.

Penning de Vries, F. W. T., Brunsting, A. H. M. & van Laar, H. H. (1974). Products, requirements and efficiency of biosynthesis: a quantitative approach. *Journal of Theoretical Biology*, **45**, 339–77.

Raven, J. A. (1976). The quantitative role of 'dark' respiratory processes in heterotropic and photolithotropic plant growth. *Annals of Botany*, **40**, 587–602.

Robson, M. J. (1982). The growth and carbon economy of selection lines of *Lolium perenne* cv. S23 with differing rates of dark respiration. I. Growth as simulated swards during a regrowth period. *Annals of Botany*, **49**, 321–9.

Ryle, G. J. A., Brockington, N. R., Powell, C. E. & Cross, B. (1973). The measurement and prediction of organ growth in a uniculm barley. *Annals of Botany*, **37**, 233–46.

Ryle, G. J. A., Cobby, J. M. & Powell, C. E. (1976). Synthetic and maintenance respiratory losses of $^{14}CO_2$ in uniculm barley and maize. *Annals of Botany*, **40**, 571–86.

Ryle, G. J. A. & Powell, C. E. (1974). The utilization of recently assimilated carbon in graminaceous plants. *Annals of Applied Biology*, **77**, 145–58.

Ryle, G. J. A., Powell, C. E. & Gordon, A. J. (1979). The respiratory costs of nitrogen fixation in soyabean, cowpea and white clover. II. Comparisons of the cost of nitrogen fixation and the utilization of combined nitrogen. *Journal of Experimental Botany*, **30**, 145–53.

Schubert, K. R. & Ryle, G. J. A. (1980). The energy requirements for nitrogen fixation in nodulated legumes. In *Advances in Legume Science*, vol. 1, ed. R. J. Summerfield & A. H. Bunting, pp. 85–96. London: Royal Botanic Gardens, Kew.

Sheehy, J. E., Cobby, J. M. & Ryle, G. J. A. (1980). The use of a model to investigate the influence of some environmental factors on the growth of perennial ryegrass. *Annals of Botany*, **46**, 343–65.

Silsbury, J. H. (1977). Energy requirement for symbiotic nitrogen fixation. *Nature, London*, **267**, 149–50.

Solomos, T. (1977). Cyanide resistant respiration in higher plants. *Annual Review of Plant Physiology*, **28**, 279–97.

Tanaka, A., Kawano, K. & Yamaguchi, J. (1966). *Photosynthesis Respiration and Plant Type of the Tropical Rice Plant*. International Rice Research Institute, Technical Bulletin 7. Los Baños, Philippines: IRRI.

Thornley, J. H. M. (1970). Respiration, growth and maintenance in plants. *Nature, London*, **227**, 304–5.

Thornley, J. H. M. (1977). Growth, maintenance and respiration: a reinterpretation. *Annals of Botany*, **41**, 1191–203.

Wilson, D. (1975). Variation in leaf respiration in relation to growth and photosynthesis of *Lolium*. *Annals of Applied Biology*, **80**, 323–38.

Wilson, D. & Jones, J. G. (1982). Effect of selection for dark respiration rate of mature leaves on crop yields of *Lolium perenne* cv. S23. *Annals of Botany*, **49**, 313–20.

Wit, C. T. de, Brouwer, R. & Penning de Vries, F. W. T. (1970). The simulation of photosynthetic systems. In *Prediction and Measurement of Photosynthetic Productivity*, ed. I. Malek, pp. 47–70. Wageningen: Centre for Agricultural Publishing and Documentation.

Yamaguchi, J. (1978). Respiration and the growth efficiency in relation to crop productivity. *Journal of the Faculty of Agriculture, Hokkaido University*, **59**, 59–129.

E. W. SIMON

2 Respiration and membrane reorganisation during imbibition

Imbibition is a remarkable phase in seed life. The mature seeds of most plant species are relatively dry and show little metabolic activity, respiration generally being so slow that it is hard to measure accurately; dry peas, for example, are reported to show 'hardly any' oxygen uptake. Nevertheless dry seeds retain a great potential for activity, which is revealed once they are supplied with water. As the seed becomes hydrated its respiration rises at a dramatic pace, until eventually if the temperature is right and the seed is not dormant, visible signs of germination and growth will ensue. The growing oxygen consumption of the young seedling must clearly be related to the production of new cells and the synthesis of enzyme systems within them (Bewley & Black, 1978).

The present account focuses, however, on the earlier period – the phase of imbibition, before there is any obvious manifestation of growth. We begin by considering the forces that cause the movement of water into imbibing seeds and then enquire into the respiratory metabolism of seeds at this stage. Evidence that the mitochondria are not fully organised and efficient in dry seed ties in with evidence that solutes leak out of imbibing seeds, for this also suggests unusual membrane properties. The final section concerns the mechanism that underlies the orderly arrangement of phospholipids and proteins in normal membranes, and how this arrangement may be modified in dry seeds.

Imbibition

Air-dry seeds commonly have water contents around 10% of their fresh weight. The water is held firmly by matric forces to cell walls or proteins, so that the water potential of a dry seed is very low indeed, about -1000 bars or less. This figure should be seen in relation to the potential of pure water, which is zero, and that of the water in a moist soil, which is only a little less as the soil solution is usually quite dilute. Because of the ensuing gradient of water potential seeds will imbibe water when it is available, at a rate depending on the magnitude of the gradient. As the seed imbibes and becomes hydrated its matric potential falls away sharply, so that its water potential rises towards that of pure water and the gradient subsides.

The rate of imbibition depends also on the resistance to flow. In soils this resistance may be considerable if the soil is relatively dry or if the seeds only make contact with the soil particles at a few small points on their surface. Petri dish experiments eliminate this factor, water being readily available, but there still remains one barrier between the water supply in the dish and the cells of the embryo within the seed. This is the testa, which may offer considerable resistance as in nuts and some legume seeds that will not normally imbibe until the seed coat is thinned down by decay or mechanical wear. Even the relatively thin testa of a pea may delay the onset of imbibition for as long as three hours (Powell & Matthews, 1978).

The progress of imbibition can be followed by weighing seeds or sectioning them to reveal the extent of the wetted region. Waggoner & Parlange (1976) have described the uptake of water by peas from which the testa had been removed to reduce variability. The entry of water into the resulting embryos is at first rapid but then slows down after about 30 minutes, presumably because water then has to travel further into the embryo tissues and encounters a greater resistance. As the water penetrates into the embryo a sharp front separates the wet and dry portions; the parts that are wetted swell up and their water content rises steadily over a period of hours.

This slow penetration of water into a seed means that although an imbibing seed may have an overall water content of 30% this is likely to be a mean value, cells at the centre still being dry while those near the surface are quite wet.

Respiration

The rapid upsurge of respiratory activity that occurs as seeds imbibe water may be the resultant of two changes: the increasing proportion of cells that have become wetted and the extent of respiratory enhancement that ensues in individual cells as they become hydrated. To assess this second factor one needs ideally some system in which the seeds can be uniformly wetted at each water content, and a method of measuring respiration sensitive enough to give reliable results with dry seeds. These conditions seem to have been met in experiments with dry seeds of charlock (*Sinapis arvensis*) which were supplied with [^{14}C]acetate in water for 24 hours at 0 °C and then redried over 40% sulphuric acid at 0 °C. Finally the seeds were left to equilibrate to known water potentials over the appropriate sulphuric acid/water mixtures for six weeks at 25 °C (Edwards, 1976). Acetate was utilised (by conversion to acetyl CoA and entry to the tricarboxylic acid cycle with release to $^{14}CO_2$) at a slow but perceptible rate even in air-dry seeds with 4–6% water;

respiratory activity increased exponentially at higher water contents. In view of the long period allowed for equilibration in these experiments, it is likely that the seed tissues became uniformly hydrated at each water potential and we may therefore conclude that the respiration of individual cells increases gradually as they imbibe water.

In beans (*Phaseolus vulgaris*) the period of rapid imbibition lasts about 16 hours, the water content of the cotyledons rising in this time from nearly 10% to just over 50%. The rate of oxygen uptake rises during imbibition, remains more or less constant for the next eight hours and then rises again without much change in water content. The enhancement of respiration during imbibition seems to be a process of physical activation rather than biochemical synthesis, for it is not much affected by change of temperature, and it is freely reversible, respiration rate ascending the curve on hydration, and descending it if the seeds are dried again (Öpik & Simon, 1963).

The respiration of seeds during the phase of imbibition proceeds with a respiratory quotient well above unity; in mung beans (Morohashi, 1978) it lies around two or three. When [^{14}C]glucose was supplied to mung beans during the first 5.5 hours of imbibition a considerable amount of radioactivity appeared in ethanol and it was calculated that 85% of the carbon dioxide produced arose by fermentation. Rather surprisingly, however, neither iodoacetate nor fluoride, both inhibitors of glycolysis, had any effect on oxygen uptake during imbibition although they inhibited carbon dioxide output by around 40%. After seven hours of imbibition, when less carbon dioxide arose by fermentation and more label appeared in malic and citric acids, iodoacetate inhibited both oxygen uptake and carbon dioxide output. Morohashi came to the conclusion that during the initial phase of imbibition oxygen uptake was limited by the capacity of the tricarboxylic acid cycle and electron transport chain, that is to say the capacity of the mitochondria. The pathway of glycolysis was fully operational and even when inhibited it still had a greater capacity than the mitochondria. Later on, when the mitochondria were no longer rate-limiting, the glycolytic inhibitors reduced oxygen uptake. Apparently the mitochondria are rather inactive at the start of imbibition and slow to reach their full potential. The same conclusion emerges from studies of labelling patterns in beans supplied with tritiated water; tricarboxylic acid cycle acids are not labelled, and so are not involved in active metabolism in the earliest periods of imbibition (Collins & Wilson, 1972).

The implications of this work can be put to the test by examining the activity of mitochondria isolated from imbibing seeds. The usual technique for preparing mitochondria from seeds involves grinding the tissue with sand in buffered medium, removing debris and centrifuging at around 25000 g to precipitate out the mitochondria. Air-dry seeds are too tough to grind in a pestle and mortar and one has therefore to powder the seeds in some kind of pulveriser before grinding. Asahi and coworkers in Japan have tried without success thus to isolate active and intact mitochondria from dry peas (Nawa and Asahi, 1973). They did obtain 'a mitochondrial pellet', but it had low malic dehydrogenase and cytochrome oxidase contents, little protein and scarcely any succinoxidase. More normal mitochondria with the expected complement of enzymes and protein, and high succinoxidase activity were obtained from peas that had been allowed to imbibe for three or better six hours. Careful analysis on sucrose gradients of 'mitochondrial' pellets prepared from dry seeds showed the presence of three cytochrome-oxidase-containing fractions: one lacking in malic dehydrogenase and high molecular weight polypeptides, one composed of relatively light fragments, and a third similar to normal mitochondria but with a different polypeptide composition.

The first two types disappeared almost completely after six hours of imbibition while the third became more active in cytochrome oxidase and malic dehydrogenase, and so more like mitochondria isolated from fully hydrated tissues. A key aspect of this maturation of mitochondria (as Asahi terms it) is the rise in protein and enzyme content. Mitochondrial maturation is unaffected by cycloheximide or chloramphenicol at concentrations sufficient to stop protein synthesis, indicating that it must involve a transfer of protein already present in the cytoplasm into the mitochondria rather than *de novo* synthesis. This is in accord with the finding that there is little or no overall increase in enzyme activity during the period of imbibition, the total amount in particles plus supernatant remaining about constant. Recently Asahi has made a detailed study of succinic dehydrogenase in pea cotyledons (Nakayama, Sugimoto & Asahi, 1980). When mitochondria were isolated from dry cotyledons by centrifuging at $20\,000\,g$, the remaining supernatant was found to contain a succinic dehydrogenase with properties very similar to those of the enzyme found in the inner membrane of mitochondria isolated from imbibed pea cotyledons. As the cotyledons imbibed water so the soluble succinic dehydrogenase disappeared. Asahi proposes that the soluble enzyme becomes associated with the inert mitochondrial membranes of dry cotyledons to yield the mature mitochondria of imbibed tissues (Fig. 1).

It may be recalled at this stage that the first step in attempting to prepare mitochondria from a dry seed powder is to grind the powder in an excess of buffered medium. Clearly this operation must allow the powder to become more or less hydrated and we have to consider why mature, active mitochondria are not produced. The answer seems to be one of time and opportunity. According to Kollöffel (1970) the oxidase activity of mitochondria prepared from pea seed is improved if the powder is first wetted for 1.5 hours. This seems to indicate that the breakage of cells in the pulveriser coupled with the dilution occasioned by the large volume of grinding medium added, separates the soluble enzymes from the immature membranes. Given time (1.5 hours) sufficient chance encounters may occur to allow assembly to proceed; but if the mixture is centrifuged straight away the components will be forcibly separated and there is no further chance of assembly. It thus so happens that the method of preparation perpetuates the condition thought to occur in the dry cells.

It may seem a little strange that mature dry cotyledons should contain relatively inert mitochondria along with the component protein molecules needed to mature them. Why are the mitochondria in dry cotyledons not already mature? It could be that the components were synthesised separately in the developing seed, ready for assembly at the start of imbibition, but if so what of the active mitochondria that must surely have been present in the developing cotyledons: have they simply disappeared? An alternative possibility is that the developing cotyledons had mature, active mitochondria which became disorganised as the cotyledons finally dried out; imbibition would then be a phase in which the disorganisation was made good, the enzymes being reassembled into the mitochondria. There is evidence to support this proposition for not only do mitochondria mature as cotyledons imbibe water, but they break down again, cytochrome oxidase and malic dehydrogenase appearing once more in the supernatant, if imbibing seeds are dried over phosphorus pentoxide (Nawa & Asahi, 1973). The same happens every year under natural conditions as seeds finally dry out when they mature on the parent plant. Thus in *Ricinus* endosperm there is a sharp fall in the activities of mitochondrial enzymes at about the time the seeds dry out, and a corresponding rise in the level of cytochrome oxidase in the supernatant (Lado, 1965).

Taken as a whole the evidence points to an assembly of proteins into mitochondria during imbibition, and the opposite, a fragmentation, during

Fig. 1. Interpretation of observations on the activity of seed mitochondria. Developing seeds and imbibing seeds yield active mitochondria, but preparations from dry seeds are relatively inactive, some enzymes remaining in the supernatant. Imbibition is a period of mitochondrial maturation when preformed enzyme protein becomes associated with membranes. This phase is reversible. If the imbibed seed is allowed to germinate, more mitochondria will appear as a result of *de novo* synthesis.

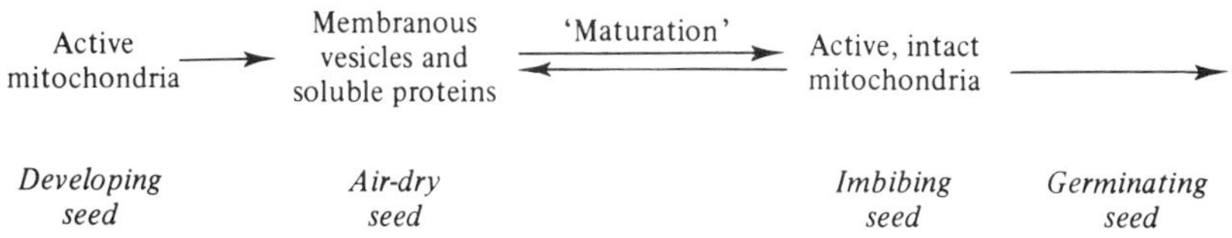

| Active mitochondria | → | Membranous vesicles and soluble proteins | ⇄ 'Maturation' | Active, intact mitochondria | → |
| Developing seed | | Air-dry seed | | Imbibing seed | Germinating seed |

desiccation (Fig. 1). Although the process of assembly is envisaged as the reverse of fragmentation it appears that the final product of imbibition may be less efficient than the original mitochondria. Thus after three imbibe/dry/imbibe cycles mitochondria from pea cotyledons have only 59% as much succinoxidase as controls imbibed only once – and they have competely lost the ability to respond to added adenosine diphosphate. It appears that errors of assembly may accumulate from one cycle to the next until ultimately the mitochondria show signs of malfunction.

Leakage

The flow of water into imbibing seeds is accompanied by a leakage of solutes in the opposite direction. This leakage first came to prominence in studies on the pre-emergence rotting of peas (Flentje & Saksena, 1964). Some batches of pea emerge poorly because many of the seeds succumb to attack in the soil by *Pythium*. The solutes that leak out of susceptible peas are capable of supporting the growth of *Pythium*, and in fact batches of pea which show low leakage in laboratory tests do best in the field. The correlation between the extent of leakage and the degree of pre-emergence mortality now forms the basis of a widely used test for seed quality.

One factor governing leakage is the state of the testa, for pea and bean seeds become more leaky if the testa is deliberately damaged with a needle or scalpel (Flentje & Saksena, 1964). The testa can be removed in its entirety from dry seeds of some pea varieties, and the resulting embryos generally leak more profusely than the original seeds (Larson, 1968). In one series of 24 hour experiments electrolyte loss from peas was from 4 to 75% greater if the testa was removed than in intact controls (Matthews & Rogerson, 1976). Seeds harvested by hand do not suffer the mechanical insults that may damage commercial seed lots (Matthews, Powell & Rogerson, 1980) – and such seeds show an eight-fold increase in leakage when the testa is removed (Powell & Matthews, 1978 and personal communication).

Several lines of evidence indicate that profuse leakage only occurs when relatively dry seeds or embryos are placed in water (Simon & RajaHarun, 1972):

1. The rate of leakage is fastest in the first minutes of imbibition and then slows down, effectively coming to a stop after 24 hours (Simon, 1978); as the seed becomes hydrated so leakage slows down. The time course of leakage follows this same pattern whether the embryos are especially prone to leak or not (Matthews & Rogerson, 1976).

2. It is possible to induce embryos to continue leakage at a relatively fast rate for several hours (albeit on an intermittent basis) by drying them over calcium chloride after each 30 minutes in water; the water content of the

embryos then never rises much and they leak rapidly each time they are returned to water.

3. Embryos prepared from peas harvested in August when they are still moist and swollen with 50% water, leak less than a tenth as much as dry peas (8% water) collected in October.

4. Experiments with seeds that have been allowed to imbibe to a limited extent from moist air or moist filter paper indicate that an overall water content of about 30–35% is critical; drier than this and leakage is progressively more rapid, while above this it is reduced to a trivial level (Simon & Wiebe, 1975).

It is unlikely that the solutes leaking away from seeds and embryos were all encrusted on the surface or present within the apoplast, for embryos transferred to fresh water at repeated intervals continue to leak (Matthews & Rogerson, 1976). On the contrary, a cytoplasmic origin is suggested by the variety of solutes that leak out, including potassium ions, organic acids, sugars and enzymes (Simon, 1974; Abdel Samad & Pearce, 1978). A major proportion of the electrolytes originally present leaks out in the course of 24 hours, about one-third being lost from intact seeds and two-thirds from isolated embryos (Simon, 1978). As the seeds or embryos were steeped in a relatively large volume of water, these figures imply that leakage ceases before the concentration of solute in the water has risen to equal that within the cells.

Two major hypotheses about the mechanism of leakage from imbibing seeds have been proposed. Larson (1968) took the view that the inrush of water by imbibing embryos was so rapid that it would force membranes aside and release cytoplasmic fluids. Matthews and his colleagues have developed this suggestion, finding support for it by tetrazolium staining. In this test hydrated seeds or embryos are placed for a few hours in a solution of tetrazolium chloride; the reagent readily enters the cells and becomes reduced to an insoluble red formazan in cells which have an active and functional dehydrogenase system. The extent and location of regions which fail to stain can be used as a guide to seed quality (Moore, 1973). Cells which fail to stain are considered dead (Powell & Matthews, 1978).

Powell & Matthews (1978) allowed pea seeds and isolated embryos to imbibe for 24 hours and then tested the cotyledons with tetrazolium. The majority of the cotyledons from the intact seeds stained completely, but the isolated embryos showed little or no staining of cells over the abaxial surface of the cotyledons. The profuse leakage from isolated embryos was thus associated with a thin layer of non-staining cells, the remaining cells in the interior of the embryos being fully stained. The non-staining surface cells would be the first to become hydrated at the start of imbibition (and it was in fact shown that a two minute exposure to water was enough to reduce the staining of these cells). The rapid initial inrush of water was thought to disrupt the organisation of the surface cells, presumably pushing membranes right out of place, and possibly causing cell death. Such a rapid flow of water would become turbulent as it was deflected by cell walls or organelles in its path and might effectively wash the solutes out of surface cells. The role of the testa in this hypothesis lies in its power to moderate the influx of water at the start of imbibition (but see Abdel Samad & Pearce, 1978; Simon, 1978). Likewise in isolated embryos the surface cells slow the rate at which water moves to the inner cells which therefore remain viable, stain with tetrazolium and so, on this argument, retain their solutes.

The total loss of solutes from a limited number of surface cells is thus held to account for the rapid initial leakage from imbibing embryos (Powell & Matthews, 1978), but it is difficult to see how it could also account for the subsequent slower phase of leakage in which 30–60% of the solutes are lost, far more than is likely to be present in the surface layers of the cotyledons: some of the solute must surely owe its origin to cells deep within the cotyledons. Some other hypothesis is needed to account for the slow, long-continued leakage that follows the initial rapid leakage from imbibing embryos, and also the slow leakage from whole seeds.

It has been proposed (Simon & RajaHarun, 1972) that cell membranes lose their integrity when seeds are air-dry, and in consequence are unable to retain

Fig. 2. Interpretation of observations on leakage from imbibing seeds. Solutes leak out of cells in the initial moments of imbibition as cell membranes are then disorganised.

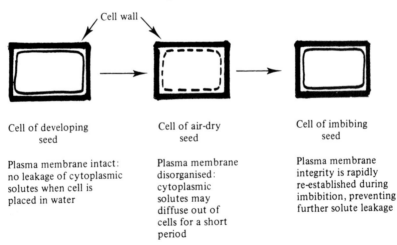

Cell of developing seed	Cell of air-dry seed	Cell of imbibing seed
Plasma membrane intact: no leakage of cytoplasmic solutes when cell is placed in water	Plasma membrane disorganised: cytoplasmic solutes may diffuse out of cells for a short period	Plasma membrane integrity is rapidly re-established during imbibition, preventing further solute leakage

cytoplasmic solutes when seeds or embryos are first placed in water. As each cell imbibes water its membranes become re-established within seconds or minutes and bring leakage to a halt (Fig. 2). Leakage from whole seeds or embryos continues for a prolonged period because the water front only penetrates slowly into seeds, and solutes released from cells deep within a cotyledon will have a long and tortuous path to traverse before they emerge into the bathing liquid. On this view the role of the testa is that of a barrier perhaps allowing the build-up of a high concentration of solute in the small space between it and the embryo. Such a concentrated solution would prevent further diffusive outflow of solutes from cotyledon cells in the short period before membrane integrity was restored and the outflow of solutes came to a complete halt.

The notion that membranes lose their integrity under dry conditions is central to the hypothesis considered here. As it would also be relevant to discussion of the state of mitochondria in dry and imbibing seeds, we must next consider what is known about membranes in dry seeds. Electron microscopy provides one approach, but unfortunately many of the earlier microscopists used aqueous fixatives which yield images that are difficult to interpret because it is not clear whether seed tissue is fixed in the dry state or whether the fixative allows more or less complete hydration before fixation. There are a number of ways round this problem such as the use of an almost entirely non-aqueous fixative (formaldehyde and glycerol) or, better, completely anhydrous fixation with osmium tetroxide vapour. Anhydrous fixation of dry rice grains shows cells that are shrunken, with highly folded walls and cell organelles with irregular outlines indicative of the shrinkage and mutual compression that must accompany the final stages of seed desiccation (Öpik, 1980). The plasma membrane is clearly recognisable and continuous – but this observation is not easy to interpret at the molecular level as there is still no certainty as to whether osmium tetroxide demarcates the position of lipids or proteins.

An alternative to cutting sections is freeze-etching, a technique in which the tissue is frozen very rapidly and then fractured with a knife to reveal a surface that can be examined by microscopy. Fractures are generally believed to occur preferentially through lipid layers, for they contain the least ice. In hydrated tissues fracture planes tend to run through the lipid-rich centre of bilayer membranes, resulting in images of the rounded surfaces of cell organelles. Cells from barley scutellum for instance, are known from thin sections to contain many protein bodies encircled by lipid droplets. These organelles can be clearly recognised in freeze-etch preparations of dry barley scutellum, provided the small block of tissue is first allowed to become hydrated (a process that is completed within two minutes). Even if the dry tissue is frozen in a drop of water so that it has only ten seconds to imbibe before all the water is frozen, the same picture emerges; the protein bodies have a rounded form and they are well spaced out. However, if portions of dry tissue are frozen in the absence of water by plunging them directly into liquid nitrogen, the image revealed is quite different, the fractures following very irregular paths mainly through scale-like bodies lying adjacent to one another in continuous arrays and identified as crushed lipid bodies (Buttrose, 1973). The fracture planes in air-dry tissue evidently pass along or through the lipid bodies rather than through membranes. At the least, it can be said that the effect of hydration on the freeze-etch image is not inconsistent with the notion that bilayer membranes become established on hydration.

Membrane architecture

The two main types of molecule in membranes are phospholipids and proteins. Phospholipids are elongated molecules based on the structure of glycerol esterified with two long-chain fatty acids; in plants they most commonly have 16 or 18 carbon atoms. The third hydroxyl of the glycerol is esterified with phosphate which may bear any one of a variety of head groups as suggested by the names of the phospholipids (phosphatidylcholine, ethanolamine, serine and inositol respectively). The salient feature of phospholipids in relation to membrane structure is that they are amphipathic, having an elongated hydrophobic tail and a more or less hydrophilic head. Ideal, dilute solutions may contain single phospholipid molecules, but in more concentrated solutions the molecules become associated with one another. They may for instance form monolayers at water surfaces, or arrange themselves in the form of micelles, spherical structures in which the tails shun the water phase, clustering together to form a central hydrophobic domain, while the hydrophilic groups lie at the surface of the sphere where they face the aqueous environment. The organisation of phospholipids in biological membranes is similar; the quantity of phospholipid present in each cell, the readiness with which some small molecules can permeate through membranes and the evidence of biophysical studies (Singer & Nicolson, 1972) are all consistent with the view that membranes are expansive areas of phospholipid bilayer.

Some of the proteins associated with membranes are readily dissociated and rendered soluble, but others are only isolated by rather drastic treatments and evidently form a part of the membrane structure. Our current concept of the location of these integral proteins in membranes is due to Singer & Nicolson (1972). The proteins are globular in nature, some of them occupying a position equivalent to a group of phospholipid molecules on one side of the bilayer while others extend right across the width of the membrane facing the aqueous region on either side. Like the phospholipids, these protein molecules owe their position in the membrane to their amphipathic nature; the elongated proteins which span the width of the membrane have hydrophilic regions at each end, with a hydrophobic central portion.

This fluid mosaic model is now widely accepted as the basis for the molecular architecture of the membranes in plant and animal cells (Fig. 3). These cells are highly hydrated and have ample water to ordain and stabilise the bilayer configuration. However, the situation in the cells of dry seeds (and other air-dry tissues) is less clear for they have such a low water potential that there cannot be any continuous aqueous phase. Is the bilayer arrangement still a realistic guide under these conditions?

Rather little direct evidence is available but some clues as to the structure of dry membranes may be gathered from physical measurements. The temperature at which heat is absorbed by a system due to a phase transition can be determined by the technique of differential scanning calorimetry. Weak phosphatidylcholine solutions show one thermal transition at the temperature at which the hydrocarbon tails 'melt', and a second one at 0 °C as the ice disappears. However, if the water content of the mixture is below about 20% then there is no thermal transition at 0 °C, indicating that all of the water is bound to the polar groups of the phospholipid. It has been calculated that about ten moles of water are bound to each mole of phosphatidylcholine; according to Chapman & Wallach (1968) the structure of a biological membrane would be severely disturbed if it was dried down beyond this.

The molecular organisation of concentrated phospholipid/water mixtures can be deduced by X-ray diffraction. It appears from work with pure

Fig. 3. Left: the fluid mosaic model of membrane structure with two integral proteins floating in the lipid bilayer. Right: hypothetical model of molecular architecture in dry seeds based on the hexagonal phase observed in phospholipid/water mixtures. The small amount of water present is located in long channels running at right-angles to the plane of the paper, and lined by the hydrophilic heads of the phospholipids and proteins. The water potential of dry seeds is so low that such an orderly array of water channels is unlikely: an irregular arrangement of slightly less dry regions is more probable.

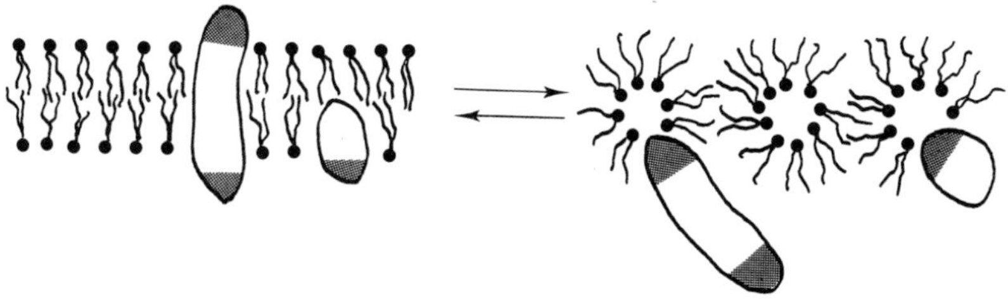

phospholipids and with crude phospholipid preparations made from a variety of tissues that a series of different molecular arrangements is possible, one hydrated phase giving way to another as the water content is reduced; this behaviour is termed lyotropic mesomorphism. Three types of lyotropic mesophase are of particular interest in connection with the likely orientation of phospholipids under relatively dry conditions. At high water contents the phospholipids in bulk phospholipid/water preparations form a stacked series of bilayers separated from one another by aqueous phases, as shown in Fig. 4. At progressively lower water contents the aqueous layers become narrower. In the second mesophase (termed hexagonal II) the molecules are arranged with their polar head groups facing long water-filled channels and their hydrocarbon tails in the dry regions between the channels (Fig. 4). This hexagonal mesophase appears in phospholipid preparations from brain tissue and from heart mitochondria, when the water content falls below about 20% (Luzzati, 1968). At water contents below about 12% at 25 °C the brain preparations (but not those from heart mitochondria) form a coagel, a phase which has been less well defined, although it is known that the hydrocarbon chains are stiff and linear.

The nature of the mesomorphic phases present in a given system depends not only on the composition of the lipid and the water content, but also on the ion content and the presence of impurities (Hauser, 1975), so that caution is needed in extrapolating from these *in vitro* studies to events inside the cells of a maturing seed that is drying out in the final stages of maturation. If maturing seeds do undergo one or more phase changes while they become desiccated, the sequence would presumably be traversed again in the opposite direction during imbibition. Transitions of this sort are likely to involve drastic alterations of permeability and could possibly account for the leakage of solutes from imbibing seeds.

Fig. 4. The arrangement of phospholipid molecules in lamellar, hexagonal and gel phases. (After Hauser, 1975.)

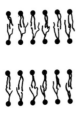

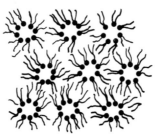

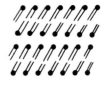

Lamellar   Hexagonal   Gel

Consider what would happen if bilayer membranes were to change over to the hexagonal arrangement at water contents below 20%. A single bilayer, such as the plasma membrane, could only generate a portion of the hexagonal phase on drying, as shown in Fig. 3, and it is enough to suggest that the reverse change during imbibition would be a period of phospholipid reorientation during which there was no effective membrane to prevent leakage. As the reorientation is a physical phenomenon it would be rapid. The fate of membrane proteins is also adumbrated in Fig. 3. During desiccation they would become displaced and although many would presumably regain their position in the membrane during imbibition, causing the mitochondria to mature, others might be lost entirely. With repeated dry/imbibe/dry cycles errors of assembly would become compounded, the mitochondria eventually losing their metabolic competency. In short, events like those depicted in Fig. 4 could account for many of the observations on leakage, mitochondrial maturation and freeze-etching.

However, McKersie & Stinson (1980) have been unable to detect a hexagonal phase in aqueous preparations of phospholipids extracted from *Lotus corniculatus* seeds. It is perhaps worth commenting that a transient hexagonal phase (like that found between 12 and 20% water in the brain phospholipid system) could have been missed in these experiments, which were confined to investigations at 40, 20, 10 and 5% moisture contents. It is also possible that phospholipid/water systems are a poor model for events in the cell. In maturing seeds membrane constituents are not exposed to pure water but to an irregular and heterogeneous environment with a declining water potential. It may well be that some neighbourhoods in the cytoplasm remain relatively moist for a longer period than others, so that the regular arrangement of phospholipids in bilayers may be dislocated on a local scale as the head groups of some molecules migrate to an adjacent region with a relatively high water potential. The final disposition of phospholipids in a dry seed might be far less regular than the arrays of channels in Figs. 3 and 4 suggest.

Conclusion

The topics reviewed in this paper and the views expressed are very similar to those put forward earlier (Simon, 1974). The proposal that membrane components are disorganised in the dry seed only to become oriented afresh into bilayers during imbibition rests on circumstantial evidence but it still seems to provide a rational basis for understanding many observations on respiratory metabolism and on leakage. It has to be admitted that we have little direct information at the molecular level about the state of membranes in dry seeds; the failure to detect a hexagonal phase in soyabean phospholipids is disappointing but may not require that the hypothesis be rejected in its entirety. Perhaps we need to investigate in model physical systems the behaviour of membrane components at water potentials as low as those found in dry seeds.

References

Abdel Samad, I. M. & Pearce, R. S. (1978). Leaching of ions, organic molecules, and enzymes from seeds of peanut (*Arachis hypogea* L.) imbibing without testas or with intact testas. *Journal of Experimental Botany*, **29**, 1471–8.

Bewley, J. D. & Black, M. (1978). *Physiology and Biochemistry of Seeds in Relation to Germination*, vol. 1. Berlin: Springer-Verlag.

Buttrose, M. S. (1973). Rapid water uptake and structural changes in imbibing seed tissues. *Protoplasma*, **77**, 111–22.

Chapman, D. & Wallach, D. F. H. (1968). Recent physical studies of phospholipids and natural membranes. In *Biological Membranes: Physical Fact and Function*, ed. D. Chapman, pp. 125–202. New York & London: Academic Press.

Collins, D. M. & Wilson, A. T. (1972). Metabolism of the axis and cotyledons of *Phaseolus vulgaris* seeds during early germination. *Phytochemistry*, **11**, 1931–5.

Edwards, M. (1976). Metabolism as a function of water potential in air-dry seeds of charlock (*Sinapis arvensis* L.). *Plant Physiology*, **58**, 237–9.

Flentje, N. T. & Saksena, H. K. (1974). Pre-emergence rotting of peas in South Australia. III. Host–pathogen interaction. *Australian Journal of Biological Sciences*, **17**, 665–75.

Hauser, H, (1975). Lipids. In *Water: A Comprehensive Treatise*, vol. 4, ed. F. Franks, pp. 209–303. New York: Plenum Press.

Kollöffel, C. (1970). Oxidative and phosphorylative activity of mitochondria from pea cotyledons during maturation of the seed. *Planta*, **91**, 321–8.

Lado, P. (1965). Inattivazione dei mitochondri negli endospermi di Ricino durante la maturazione. *Giornale Botanico Italiano*, **72**, 359–69.

Larson, L. A. (1968). The effect soaking pea seeds with or without seedcoats has on seedling growth. *Plant Physiology*, **43**, 255–9.

Luzzati, V, (1968). X-ray diffraction studies of lipid–water systems. In *Biological Membranes: Physical Fact and Function*, ed. D. Chapman, pp. 71–123. New York & London: Academic Press.

McKersie, B. D. & Stinson, R. H. (1980). Effect of dehydration on leakage and membrane structure in *Lotus corniculatus* L. seeds. *Plant Physiology*, **66**, 316–20.

Matthews, S., Powell, A. A. & Rogerson, N. E. (1980). Physiological aspects of the development and storage of pea seeds and their significance to seed production. In *Seed Production*, ed. P. B. Hebblethwaite, pp. 513–25. London: Butterworth.

Matthews, S. & Rogerson, N. E. (1976). The influence of embryo condition on the leaching of solutes from pea embryos. *Journal of Experimental Botany*, **27**, 961–8.

Moore, R. P. (1973). Tetrazolium staining for assessing seed quality. In *Seed Ecology*, ed. W. Heydecker, pp. 347–66. London: Butterworth.

Morohashi, Y. (1978). Development of respiratory metabolism in seeds during hydration. In *Dry Biological Systems*, ed. J. H. Crowe & J. S. Clegg, pp. 225–40. New York & London: Academic Press.

Nakayama, N., Sugimoto, I. & Asahi, T. (1980). Presence in dry pea cotyledons of soluble succinate dehydrogenase that is assembled into the mitochondrial inner membrane during seed imbibition. *Plant Physiology*, **65**, 229–33.

Nawa, Y. & Asahi, T. (1973). Relationship between the water content of pea cotyledons and mitochondrial development during the early stage of germination. *Plant and Cell Physiology*, **14**, 607–10.

Öpik, H. (1980). The ultrastructure of coleoptile cells in dry rice (*Oryza sativa* L.) grains after anhydrous fixation with osmium tetroxide vapour. *New Phytologist*, **85**, 521–9.

Öpik, H. & Simon, E. W. (1963). Water content and respiration rate of bean cotyledons. *Journal of Experimental Botany*, **14**, 299–310.

Powell, A. A. & Matthews, S. (1978). The damaging effect of water on dry pea embryos during imbibition. *Journal of Experimental Botany*, **29**, 1215–29.

Simon, E. W. (1974). Phospholipids and plant membrane permeability. *New Phytologist*, **73**, 377–420.

Simon, E. W. (1978). Membranes in dry and imbibing seeds. In *Dry Biological Systems*, ed. J. H. Crowe & J. S. Clegg, pp. 205–24. New York & London: Academic Press.

Simon, E. W. & RajaHarun, R. M. (1972). Leakage during seed imbibition. *Journal of Experimental Botany*, **23**, 1076–85.

Simon, E. W. & Wiebe, H. H. (1975). Leakage during imbibition, resistance to damage at low temperature and the water content of peas. *New Phytologist*, **74**, 407–11.

Singer, S. J. & Nicolson, G. L. (1972). The fluid mosaic model of the structure of cell membranes. *Science*, **175**, 720–31.

Waggoner, P. E. & Parlange, J.-Y. (1976). Water uptake and water diffusivity of seeds. *Plant Physiology*, **57**, 153–6.

CHAIM FRENKEL

3 Factors regulating the respiratory upsurge in developing storage tissues

Changes in respiratory patterns in developing plant storage organs

The growth and development of plant storage organs (e.g. fruit, tubers) is occasionally accompanied by a respiratory burst. The changes in fruit respiration serve to illustrate the point. Fig. 1 shows that following a decline in respiratory gas exchange during the rapid growth phase, respiration reaches a steady state by the time the fruit has entered the stage of maturation. The English workers Kidd & West (1925, 1930), among the first to study fruit respiration, observed that maturing apple fruit exhibits a respiratory upsurge in conjunction with the onset of ripening; they termed the respiratory burst the 'climacteric' respiration. Additional work revealed that pome and other fruit, notably tropical and subtropical crops, exhibit the climacteric respiration. Others, including strawberry, cherries, melons, figs, pineapple, and an assortment of different berries, show a continuous downward drift in respiration throughout the stage of maturation and ripening, and have been classified by Biale (1960, 1964) as non-climacteric.

Fig. 1. Changes in respiratory patterns of climacteric and non-climacteric fruit in relation to fruit development. (From Biale, 1964.)

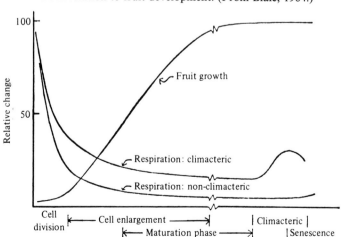

It now appears that fruits behave as climacteric or non-climacteric according to the ability of tissues to synthesise and respond to ethylene, either endogenous or applied (McGlasson, 1970; Rhodes, 1970). In climacteric fruit the respiratory upsurge is accompanied, and often preceded, by an increase in the autocatalytic synthesis and evolution of ethylene, and it is the rise in endogenous ethylene which apparently leads to the respiratory upsurge in these fruit. By comparison, in non-climacteric fruit ethylene concentrations are low and continue a downward drift; this trend is consequently accompanied by a similar decline in respiration. The two tissue classes differ also in their responses to applied ethylene (Fig. 2). In climacteric fruit ethylene initiates its own autocatalytic synthesis and subsequently the emergence of the respiratory upsurge. Once it has been initiated, and completed its course, the climacteric respiration cannot be reproduced. In contrast, in non-climacteric fruit applied ethylene does not lead to autocatalytic synthesis. The elicited respiratory upsurge can often be reproduced by repeated applications of the olefin, and the magnitude of the response is a function of the concentration of the applied ethylene, whereas in climacteric fruit it is not (Fig. 2). Many storage organs, e.g. sweet or white potato tubers, behave like non-climacteric tissue (Reid & Pratt, 1972; Chin & Frenkel, 1977; Day, Arron & Laties, 1980).

Although ethylene metabolism and action in these tissues has received wide attention it appears that other compounds have similar effects. For example,

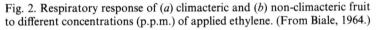

Fig. 2. Respiratory response of (*a*) climacteric and (*b*) non-climacteric fruit to different concentrations (p.p.m.) of applied ethylene. (From Biale, 1964.)

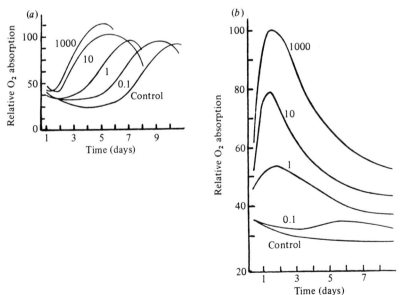

other naturally occurring volatiles, including ethanol and acetaldehyde, are produced in considerable quantities in ripening fruit (Smagula & Bramlage, 1977; Janes & Frenkel, 1978) and stressed potato tubers (Smagula & Bramlage, 1977; Varns & Glynn, 1979), and the application of these

Fig. 3. Respiratory upsurge in potato tubers following the application of different concentrations (p.p.m.) of acetaldehyde, ethanol or acetic acid in air or in 100% oxygen.

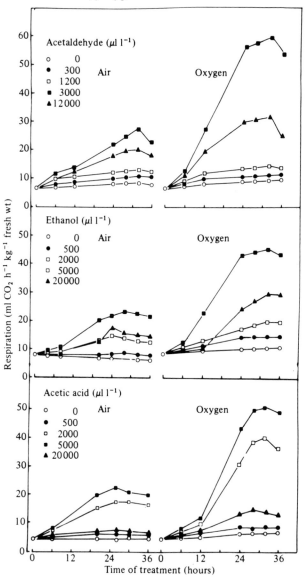

compounds can elicit in potatoes respiratory changes resembling those caused by ethylene, including stimulation of the respiratory upsurge (Fig. 3), the turnover of respiratory intermediates (Rychter *et al.*, 1979*b*) and a shift in respiratory paths, i.e. development of cyanide-resistant respiration (Rychter, Janes & Frenkel, 1979*a*). Applied carbon dioxide can also trigger similar responses in some non-climacteric storage tissues including carrots (Weichmann, 1973) and potatoes (Perez-Trejo, Janes & Frenkel, 1981), whereas in climacteric fruit carbon dioxide is known as an inhibitor of respiration (Young, Romani & Biale, 1962). A respiratory upsurge can be induced also by chilling temperatures, usually below 12 °C. For example, the late-ripening pear varieties Anjou and Bosc show a post-harvest resistance to ripening, including the onset of the climacteric respiration; this resistance can be overcome by applied ethylene but also by holding the fruit at chilling temperature (Porritt, 1965). In fruit the low-temperature treatment may represent an indirect ethylene effect, since it stimulates the synthesis of ethylene (Sfakiotakis & Dilley, 1974), but a similar respiratory response can also be observed in chilled potato tubers (Shulze *et al.*, 1979), which do not synthesise ethylene in sufficient quantities to trigger a respiratory upsurge (Reid & Pratt, 1972; Rychter *et al.*, 1979*b*). It is, therefore, likely that the effect of chilling temperature in the fruit is also independent of ethylene action.

Collectively, the information outlined above indicates that the stimulation of respiratory activity may be a response to a variety of factors. In ripening fruit ethylene may be predominant but other volatiles may also contribute to the onset of the respiratory response. Anomalous conditions including chilling temperatures, and the emanation of volatiles including carbon dioxide, ethanol and acetaldehyde from diseased tissues, may trigger an increase in respiration in non-climacteric tissues.

The metabolic nature of the respiratory upsurge

Since its discovery by the English workers (Kidd & West, 1925, 1930; Blackman & Parija, 1928) the respiratory climacteric has been the focus of attention, as a process signifying an important transition in tissue metabolism, but also as a source of controversy. Blackman & Parija (1928) were the early proponents of the concept that this respiratory response represented a loss of 'organisation resistance'. The concept envisioned biological membranes as the basis for cellular organisation – maintaining reactants, including enzymes and their corresponding substrates, sequestered in their appropriate cellular compartments, and exerting control by a stepwise release of reacting substances. Loss of membrane structures and function, and thereby of organisation resistance, establishes the conditions for reactions, including respiration, to occur at an unmitigated rate: hence the onset of the respiratory upsurge. More recent biochemical and microscopical evidence indicates that the respiratory upsurge in fruit is in fact associated with some losses in membrane structure (Bain & Mercer, 1964) and function (Sacher, 1962). However, membrane degradation, as shown for example in fruit chloroplasts (Bain & Mercer, 1964), is a limited phenomenon and moreover may be coincidental to other primary process(es). In ripening fruit it has been clearly delineated from the respiratory upsurge (Palmer, 1970; Rhodes, 1970).

The concept that the respiratory upsurge represents loss of metabolic control was extended specifically to the study of fruit mitochondria. Millerd, Bonner & Biale (1953) examined the possibility that the electron flux from substrate to molecular oxygen in the mitochondrial cytochrome pathway is relieved from restriction by the coupled process of phosphorylation. Presumably in ripening fruit tissue the activity of naturally occurring phenolic or other substances serves to uncouple the electron flow from the phosphorylation of ADP. Thereby, the rate of transfer of reducing equivalents from substrate to oxygen may become a function of the inherent capacity of the respiratory chain to catalyse the process. This concept is in conflict, however, with evidence showing that the respiratory rise is accompanied by the formation of ATP, as occurs for example in senescing leaves (Manicol, Young & Biale, 1973; Malik & Thimann, 1980), ripening fruit (Young & Biale, 1967; Solomos & Biale, 1974; Solomos & Laties, 1976) or ethylene-treated potato tubers (Solomos & Laties, 1975). This evidence suggests that the phosphorylating capacity of mitochondria is not impaired, and on the contrary, the organelles may be functioning at or near full capacity in the production of ATP. This conclusion is further supported by studies showing that mitochondrial ultrastructure and function are preserved throughout the ripening process (Bain & Mercer, 1964).

Since the structure and function of mitochondria are kept intact one needs to consider how the organelles can engage in extensive oxygen uptake even though the adenylate balance, particularly the decrease in the availability of ADP, may restrict coupled respiration (Atkinson, 1969). To account for this discrepancy it was suggested that in actively respiring tissues mitochondria develop an alternative pathway which can circumvent respiratory control (Day *et al.*, 1980). Fig. 4 (see also Palmer, this volume), outlining the relationship of this path to the overall electron transfer scheme from substrate to oxygen, indicates that the pathway accounts for a single phosphorylation site (or none when succinate is used as substrate). The alternative path, consisting of an as yet undefined terminal oxidase, is insensitive to cyanide, azide and other metabolic inhibitors of cytochrome oxidase (hence it is often termed the cyanide-resistant respiration). The rate of respiration catalysed by the alternative pathway may be a function of substrate availability, the inherent capacity for electron transfer, and the rate of ADP regeneration. In the event that the availability of ADP becomes limiting, the single phosphorylation site may allow electron transfer via the alternative path at three times the rate permitted by the cytochrome path. There is a considerable body of evidence showing that the respiratory upsurge in storage tissues is associated with the development of cyanide-resistant respiration – for example, in ripening climacteric fruit (Day *et al.*, 1980), or in potatoes treated with ethylene (Rychter *et al.*, 1979*a*), acetaldehyde or ethanol (Janes, Rychter & Frenkel, 1979) or by chilling temperatures (Shulze *et al.*, 1979). The major difficulty with this concept is firstly the need to demonstrate that the pathway is engaged and operational in these tissues, mediating electron transfer to oxygen, and in addition that the apportioning of electrons to the cyanide-resistant pathway is responsible for the respiratory upsurge (Day *et al.*, 1980). In the absence of direct and compelling evidence the emergence of cyanide-resistant respiration may perhaps be regarded as coincidental to the onset of the respiratory upsurge.

Another possibility which may account for a respiratory upsurge in excess of the mitochondrial capacity is oxygen uptake by extramitochondrial terminal oxidases. These may include microsomal oxidases such as cytochrome P-450 (Reichhart *et al.*, 1980), cytoplasmic enzymes such as xanthine oxidase or aldehyde oxidase (Rothe, 1974; Fridovich, 1975), oxidases found in microbodies including glycolate oxidase or amino acid oxidases (Tolbert, 1971), and apparently cell wall enzymes (Elstner & Heupel, 1976). The evidence that any of these systems may contribute to the respiratory upsurge is at present circumstantial. For example, when the activities of both the cytochrome and the cyanide-insensitive oxidases are concurrently blocked

Fig. 4. Proposed electron pathway in plant mitochondria involving the alternative (cyanide-resistant) oxidase in relation to the cytochrome pathway. Sites of inhibitor action are indicated by bars. Sites of ATP production are indicated by roman numerals. Fe-S, iron sulphur protein; Fp, flavoprotein; Q, ubiquinone; SHAM, salicyl hydroxamic acid. (From Day *et al.*, 1980.)

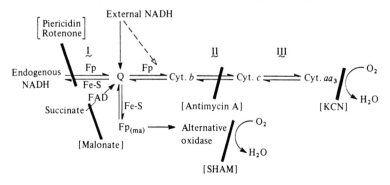

with the use of the appropriate inhibitors (Fig. 4) there is invariably a persistent residual respiration of an unidentified nature (Rychter *et al.*, 1979*a*; Day *et al.*, 1980), and in some instances (e.g. carbon-dioxide-treated potato tubers) the magnitude of this type of respiration may account for almost half of the consumed oxygen (Table 1).

Respiratory products may also be used to obtain a clue as to the possible identity of terminal oxidases contributing to the respiratory upsurge. While water is the common mitochondrial respiratory product, resulting from the tetravalent reduction of oxygen, the univalent reduction of oxygen may result in partially reduced oxygen intermediates, including hydrogen peroxide. In potatoes the ethylene-induced respiratory upsurge appears to be associated with the onset of hydrogen peroxide production (Fig. 5), and similar changes in hydrogen peroxide, closely resembling the climacteric respiration, were observed in pear (Brennan & Frenkel, 1977) and tomato (Frenkel & Eskin, 1977). The oxidase(s) representing this type of respiration exhibit low affinity for oxygen (Tolbert, 1971), and while the mitochondrial cytochrome oxidase is operative it may effectively compete for the limited oxygen supply in bulky storage tissues and thereby restrict the activity of other oxidases. Under high oxygen tensions the accessibility to oxygen may not be limiting. In fact, when a respiratory upsurge is induced in potatoes by the application of ethylene, or other volatiles, coupling of these treatments with 100% oxygen markedly enhances the magnitude of the respiratory response (Figs. 3, 5), and results in the accompanying production of hydrogen peroxide. These results are in keeping with the concept that the respiratory upsurge may reflect, in part, the activity of oxidases exhibiting low affinity for oxygen and catalysing the production of partially reduced oxygen forms. These are features which are clearly distinguishable from mitochondrial respiration since the activity of the latter can persist at relatively low oxygen tensions and is not accompanied

Table 1. *Effect of respiratory inhibitors (potassium cyanide, SHAM^a) on the uptake of oxygen by freshly cut slices from potato tubers held in air plus 10% carbon dioxide for 0, 24, 48 and 72 hours*

Hours in CO_2 atmosphere	O_2 uptake (% of control without inhibitors)		
	Control	KCN	KCN+SHAM
0	100	10	7
24	100	16	12
48	100	55	20
72	100	98	40

^{a} SHAM, salicyl hydroxamic acid.

Fig. 5. Respiratory upsurge (*a*, *b*) and concomitant changes in hydrogen peroxide content (*c*, *d*) in potato tubers following the application of ethylene (open triangles) or the ethylene analogue propylene (filled circles).

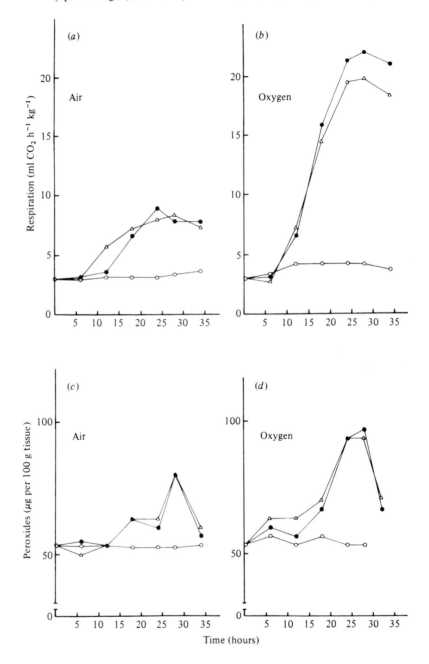

by the production of substantial levels of hydrogen peroxide, even when the alternative cyanide-resistant path has developed in the organelle (Day *et al.*, 1980).

It appears that a complex system consisting of terminal oxidases may catalyse the gas exchange associated with the respiratory upsurge. Given that the proposed concept is valid, it is desirable to identify the respiratory centres and quantify their contribution to the total respiratory gas exchange. Low affinity for molecular oxygen and production of partially reduced oxygen species may be used as experimental probes in the identification of these oxidase(s).

Metabolic role of the respiratory upsurge

One of the obvious functions of respiration is the release of the energy which is inherent in the structure of the respiratory substrates. Accordingly, an increase in ATP production accompanies the respiratory upsurge in ripening fruit (Young & Biale, 1967; Solomos & Biale, 1974; Solomos & Laties, 1976), ethylene-treated potatoes (Solomos & Laties, 1975) or senescing leaves (Malik & Thimann, 1980).

Production of copious levels of ATP appears, however, to present an anomaly from the stand-point of energy requirement and management. Changes in ATP levels, together with those of ADP and AMP, serve to establish in tissues adenylate ratios – defined by Atkinson (1969) as the energy charge – which function to regulate and synchronise the utilisation of substrates in respiration and glycolysis. Under anaerobic conditions the decline in ATP produced in respiration and corresponding changes in the energy charge cause the acceleration in glycolysis by allosteric regulation. Accelerated rates of glycolysis were inferred, however, under aerobic conditions in respiring banana (Barker & Solomos, 1962), tomato (Chalmers & Rowan, 1971) and avocado (Young & Biale, 1967; Solomos & Biale, 1974), and in conjunction with the respiratory upsurge in potatoes treated with ethylene, hydrogen cyanide (Solomos & Laties, 1975), acetaldehyde or ethanol (Rychter *et al.*, 1979*b*). Whereas acceleration of glycolysis during anaerobic conditions may result from the decline in the energy charge, this explanation may not apply to rapidly respiring storage tissues since aerobic respiration is associated with extensive phosphorylative activity and an increase in the energy output. Thus, while respiration will tend to curtail glycolysis, presumably by competing for ADP and the production of ATP at levels which are inhibitory to glycolysis, in reality the stimulation in respiration appears to be associated with *enhanced* rates of glycolysis. These conditions suggest that glycolysis has become disengaged from regulation by the energy charge associated with the respiratory upsurge.

It is generally agreed that the ATP levels produced during normal respiratory activity are sufficient to satisfy the metabolic needs of developing storage tissues, and there is no obvious reason why respiration and the concurrent production of ATP should be dramatically increased (Solomos & Biale, 1974). Instead, in actively respiring storage tissues the utilisation of limited metabolic resources for the production of energy is accelerated for no apparent reason and appears incongruous with the metabolic requirements and efficient energy management. These are some of the considerations suggesting that the metabolic role of the respiratory upsurge may not be dictated by the energy metabolism in storage tissues and may lie elsewhere.

It stands to reason that when glycolysis and respiration are disengaged from regulation by the energy charge, and possibly other factors, substrate availability may become the rate-limiting factor in these processes. The release of glucose from storage carbohydrates is clearly a major source of substrate which accompanies the onset of respiration; this is shown, for example, by the coincidence of starch utilisation and the onset of climacteric respiration in ripening avocado (Pesis, Fuchs & Zanberman, 1978) or carbon-dioxide-treated potatoes (Perez-Trejo et al., 1981). Consumption of other substrates in conjunction with the respiratory upsurge includes the utilisation of organic acids (Young et al., 1962) and possibly other less obvious sources of respiratory intermediates. Release of respiratory substrates which were previously kept compartmentalised or in inactive form (e.g. storage carbohydrates) may lead to a state of toxicity, however. Excess glucose is believed to cause modification in the structure and function of proteins, some of which may be critical to cellular function (Sharon, 1980). Release of compartmentalised amino or organic acids may lead to pH changes in tissues, with detrimental consequences to a host of metabolic paths (Davies, 1973). Analogous damaging effects could be elicited by the release into the cytosolic environment of other endogenous compounds, notably phenolics. While several metabolic alternatives are available to neutralise these compounds, detoxification by oxidative degradation is a reasonable solution for this predicament. Glycolysis and mitochondrial respiration are logical outlets for the utilisation of released sugars, organic, fatty and amino acids, and many other secondary metabolites. The activation of the microsomal enzymes, including cytochrome P-450 and the associated hydroxylating systems, as a mechanism for altering the activity of endogenous or applied substances in plants (Reichhart et al., 1980) epitomises the concept that oxidative metabolism may be mobilised as a detoxification response. There are indications, as outlined above, that other oxidative pathways may also be activated apparently for similar reasons.

Similarly, carbon dioxide evolution may represent the collective output of several decarboxylating mechanisms. Most of the evolved carbon dioxide undoubtedly reflects substrate breakdown as mediated by the tricarboxylic acid cycle during mitochondrial respiration. Other paths, for example the degradation of malic acid by malic enzymes or the decarboxylation of pyruvate to acetaldehyde (Ulrich, 1970), may also contribute to carbon dioxide output. The hydrogen peroxide produced during the respiratory upsurge can spontaneously degrade glyoxalate and other alpha-keto acids (Elstner & Heupel, 1976) with the concomitant release of carbon dioxide.

It may not be a coincidence that the respiratory upsurge in storage tissue, notably fruit, is accompanied by the formation of inordinate by-products. The profuse evolution of aroma compounds, originating for the most part from the oxidative degradation of lipids, or storage amino and organic acids (Eskin, 1979), may be a related phenomenon arising from the mobilisation of respiratory metabolism to detoxify released substances. Similarly, the copious emanation of volatile amines in skunk cabbage spadix may be associated with the spectacular respiratory upsurge in the plant (Day *et al.*, 1980).

The concept that the mass release of metabolites, mainly from membranous compartments, triggers the respiratory upsurge is at present largely discounted, in favour of the view that membrane leakiness may be restricted and selective (McGlasson, 1970; Palmer, 1970; Rhodes, 1970) and, moreover, a regulated process as indicated in senescing flower petals (Suttle & Kende, 1980). The contemporary state of our knowledge and understanding has not resolved, however, the lingering controversy between the views of Kidd & West (1925, 1930) and the opposing concept of Blackman. Namely, is the respiratory upsurge in storage tissue including fruit, an outcome of regulated metabolic events, i.e. enzyme synthesis or activation, or is the increase in the activity of respiratory enzymes a response to the mass release of substrates as a consequence of the metabolic deregulation of tissues.

References

Atkinson, D. F. (1969). Regulation of enzyme function. *Annual Review of Microbiology*, **23**, 46–68.

Bain, J. M. & Mercer, F. V. (1964). Organization resistance and the respiration climacteric. *Australian Journal of Biological Sciences*, **17**, 78–85.

Barker, J. & Solomos, T. (1962). The mechanism of the climacteric rise in respiration in banana fruits. *Nature, London*, **109**, 189–91.

Biale, J. B. (1960). Respiration of fruits. *Encyclopedia of Plant Physiology*, **12**, 537–92.

Biale, J. B. (1964). Growth, maturation, and senescence in fruits. *Science*, **146**, 880–8.

Blackman, F. F. & Parija, P. (1928). Analytical studies in plant respiration. I. The respiration of a population of senescent ripening apples. *Proceedings of the Royal Society of London, Series B*, **103**, 412–45.

Brennan, T. & Frenkel, C. (1977). Involvement of hydrogen peroxide in the regulation of senescence in pear. *Plant Physiology*, **59**, 411–16.

Chalmers, D. J. & Rowan, S. (1971). The climacteric in ripening tomato fruit. *Plant Physiology*, **48**, 235–40.

Chin, C. & Frenkel, C. (1977). Upsurge in respiration and peroxide formation in potato tubers as influenced by ethylene, propylene, and cyanide. *Plant Physiology*, **59**, 515–18.

Davies, D. D. (1973). Control of and by pH. *Symposia of the Society for Experimental Biology*, **27**, 513–29.

Day, D. A., Arron, G. P. & Laties, G. G. (1980). Nature and control of respiratory pathway in plants: the interaction of cyanide resistant respiration with the cyanide sensitive pathway. In *The Biochemistry of Plants. A Comprehensive Treatise*, vol. 1, ed. D. D. Davies. New York & London: Academic Press.

Elstner, E. F. & Heupel, A. (1976). Formation of hydrogen peroxide by isolated cell walls from horseradish. *Planta*, **130**, 175–80.

Eskin, N. A. M. (1979). Aldehydes, alcohols and esters: biogenesis. In *Plant Pigments, Flavors and Textures: The Chemistry and Biochemistry of Selected Compounds*, ed. N. A. M. Eskin. New York & London: Academic Press.

Frenkel, C. & Eskin, N. A. M. (1977). Ethylene evolution as related to changes in hydroperoxides in ripening tomato fruit. *HortScience*, **12**, 552–3.

Fridovich, I. (1975). Superoxide dismutase. *Annual Review of Biochemistry*, **44**, 147–59.

Janes, H. W. & Frenkel, C. (1978). Promotion of ripening processes in pear by acetaldehyde, independently of ethylene action. *Journal of the American Society of Horticultural Science*, **103**, 397–400.

Janes, H. W., Rychter, A. & Frenkel, C. (1979). Factors influencing the development of cyanide resistant respiration in potato tissue. *Plant Physiology*, **63**, 837–40.

Kidd, F. & West, C. (1925). *The Course of Respiratory Activity throughout the Life of an Apple*. Great Britain Department of Scientific and Industrial Research, Food Investigation Report 1924, pp. 27–33.

Kidd, F. & West, C. (1930). Physiology of fruit. I. Changes in respiratory activity of apples during their senescence at different temperatures. *Proceedings of the Royal Society of London, Series B*, **106**, 93–109.

McGlasson, W. B. (1970). The ethylene factor. In *The Biochemistry of Fruits and their Products*, ed. A. C. Hulme. New York & London: Academic Press.

Malik, N. S. A. & Thimann, R. V. (1980). Metabolism of oat leaves during senescence. IV. Changes in ATP levels. *Plant Physiology*, **65**, 855–8.

Manicol, P. K., Young, R. E. & Biale, J. B. (1973). Metabolic regulation in the senescing tobacco leaf. I. Changes in pattern of ^{32}P incorporation into leaf disc metabolites. *Plant Physiology*, **51**, 793–7.

Millerd, A., Bonner, J. & Biale, J. B. (1953). The climacteric rise in respiration as controlled by phosphorylative coupling. *Plant Physiology,* **28**, 521–31.

Palmer, J. R. (1970). The banana. In *The Biochemistry of Fruits and their Products,* ed. A. C. Hulme. New York & London: Academic Press.

Perez-Trejo, M. S., Janes, H. W. & Frenkel, C. (1981). Mobilization of respiratory metabolism in potato tubers by carbon dioxide. *Plant Physiology,* **67**, 514–17.

Pesis, E., Fuchs, Y. & Zanberman, G. (1978). Starch content and amylase activity in avocado fruit. *Journal of the American Society of Horticultural Science,* **103**, 673–6.

Porritt, S. W. (1965). The effect of temperature on post-harvest physiology and storage life of pears. *Canadian Journal of Plant Science,* **44**, 568–79.

Reichhart, D., Salaun, J. P., Benveniste, I. & Durst, F. (1980). Time course induction of cytochrome P-450, NADPH-cytochrome *c* reductase, and cinnamic acid hydroxylase by phenobarbital, ethanol, herbicides, and manganese in higher plant microsomes. *Plant Physiology,* **66**, 600–4.

Reid, S. M. & Pratt, H. K. (1972). Effect of ethylene on potato tuber respiration. *Plant Physiology,* **49**, 252–5.

Rhodes, M. J. C. (1970). The climacteric and ripening of fruit. In *The Biochemistry of Fruits and their Products,* ed. A. C. Hulme. New York & London: Academic Press.

Rothe, G. M. (1974). Aldehyde oxidase isozymes in potato tubers. *Plant and Cell Physiology,* **15**, 493–9.

Rychter, A., Janes, H. W. & Frenkel, C. (1979*a*). Effect of ethylene and oxygen on the development of cyanide resistant respiration in whole potato mitochondria. *Plant Physiology,* **63**, 149–51.

Rychter, A., Janes, H. W., Chin, C. & Frenkel, C. (1979*b*). The effect of ethylene and end products of glycolysis on changes in respiration and respiratory metabolites in potato tubers. *Plant Physiology,* **64**, 108–11.

Sacher, J. A. (1962). Relations between changes in membrane permeability and the climacteric in banana and avocado. *Nature, London,* **195**, 577–8.

Sfakiotakis, E. M. & Dilley, D. R. (1974). Induction of ethylene production in Bosc pears by post-harvest cold stress. *HortScience,* **9**, 336–8.

Sharon, N. (1980). Carbohydrates. *Scientific American,* **243**, 90–116.

Shulze, C., Wulster, G., Janes, H. W. & Frenkel, C. (1979). Interaction of low temperature and oxygen regimes on the stimulation of respiration in potato tubers. *Plant Physiology,* **63**, S572.

Smagula, J. M. & Bramlage, W. J. (1977). Acetaldehyde accumulation: is it a cause of physiological deterioration of fruits? *HortScience,* **12**, 200–3.

Solomos, T. & Biale, J. B. (1974). Respiration and fruit ripening. In *Facteurs et regulation de la maturation des fruits.* Colloques Internationaux de CNRS No. 238. Paris.

Solomos, T. & Laties, G. G. (1975). The mechanism of ethylene and cyanide action in triggering the rise in respiration in potato tubers. *Plant Physiology,* **55**, 73–8.

Solomos, T. & Laties, G. G. (1976). Effect of cyanide and ethylene on the respiration of cyanide-sensitive and cyanide-resistant plant tissues. *Plant Physiology,* **58**, 47–50.

Suttle, J. C. & Kende, H. (1980). Ethylene action and loss of membrane integrity during petal senescence in *Tradescantia*. *Plant Physiology*, **65**, 1067–72.

Tolbert, N. W. (1971). Microbodies, peroxisomes, glyoxysomes. *Annual Review of Plant Physiology*, **22**, 45–74.

Ulrich, R. (1970). Organic acids. In *The Biochemistry of Fruits and their Products*, ed. A. C. Hulme. New York & London: Academic Press.

Varns, J. L. & Glynn, M. T. (1979). Detection of disease in stored potatoes by volatile monitoring. *American Potato Journal*, **57**, 185–97.

Weichmann, J. (1973). Die Wirkung unterschiedlichen CO_2 Partialdruckes auf den Gasstoffwechsel von Möhren. *Gartenbauwissenschaft*, **38**, 248–52.

Young, R. E. & Biale, J. B. (1967). Phosphorylation in avocado fruit slices in relation to the respiratory climacteric. *Plant Physiology*, **42**, 1357–62.

Young, R. E., Romani, R. J. & Biale, J. B. (1962). Carbon dioxide effects on fruit respiration. *Plant Physiology*, **37**, 416–22.

BASTIAAN J. D. MEEUSE

4 Physiological and biochemical aspects of thermogenic respiration in the aroid appendix

It is now more than two centuries ago that Jean Baptiste de Monet, Chevalier de Lamarck, noticed the spectacular temperature rise displayed by the inflorescence of *Arum italicum*, which he treated as a variety of *Arum maculatum*. Since then, many other instances of thermogenicity in arum lilies have been described (see Meeuse, 1975, for a review). The phenomenon is displayed also by certain palms, cycads, water lilies and Pandanaceae (see below).

Recently, a number of review articles have appeared dealing with the events that take place in the mitochondria, which are involved in the respiratory processes that give rise to the heat production (Palmer, 1976; Day, Arron & Laties, 1980; Storey, 1980; Laties, 1982). Unfortunately, however, a wide gap still exists between investigators following the subcellular approach to thermogenicity on the one hand and, on the other hand, 'classical' physiologists and those floral biologists concerned with the ecological and evolutionary significance of the phenomenon. The present article represents an attempt to narrow the gap.

Terminology

In the Aroideae, a taxon which includes genera such as *Arum*, *Sauromatum* (Fig. 1) and *Dracunculus*, the inflorescences show a very high degree of functional differentiation. The fleshy central column or spadix of the inflorescence is covered, at its base, by a dense group of strongly reduced pistillate ('female') flowers. The staminate ('male') flowers, likewise strongly reduced so that they essentially form sessile stamens, are bunched together higher up on the spadix. The latter is surrounded by a large bract, the spathe, the base of which is transformed into a closed floral chamber. When the inflorescence reaches maturity, the spathe unfolds, exposing the naked, sterile upper part of the spadix known as the appendix. In addition to being capable of producing a great deal of heat, at least in some species, the appendix develops a strong odour; for this reason Vogel (1963) refers to it as an osmophore. The heat serves as a volatiliser for the odour, which in turn acts as the pollinator-attractant.

In *Arum maculatum* a group of stiff bristles (barrier organs; modified flowers) is found above the staminate flowers, blocking the entrance (or exit) of the floral chamber. These act as a sieve or colander, keeping out large

Fig. 1. The inflorescence of *Sauromatum guttatum*, the voodoo lily. Left: the inflorescence emerges from the tuber or corm without the necessity for any water or soil – the reason why this plant is such an excellent experimental guinea-pig. Right: lengthwise section through the floral chamber to show the arrangement of the various organs.

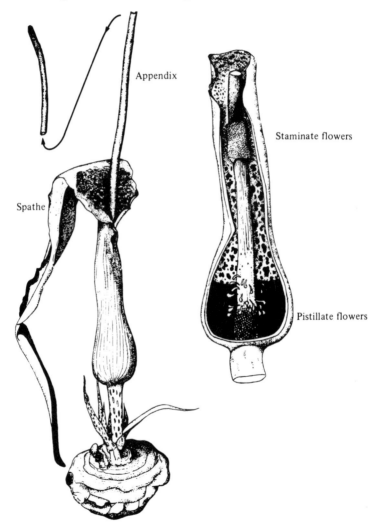

insects such as bluebottle or greenbottle flies which might be attracted by the smell of the appendix, but allowing the 'legitimate' pollinators (small psychodid flies in the case of *Arum maculatum*, small beetles in that of *A. nigrum*) to fall through. Escape from the floral chamber is impossible for them, because of the barrier organs and the fact that the chamber walls also have oil droplets on their surface. The inflorescences display a very pronounced protogyny: when the appendix is active in producing heat and smell, and the insects are arriving in the floral chamber, the pistillate flowers at the base of the spadix are receptive, that is, ready to be pollinated, while the staminate flowers (anthers) higher up are still firmly closed and are not shedding their pollen. The visitors find food, which the pistillate flowers offer them in the form of a stigmatic secretion; since some of the animals may be carrying pollen obtained from another *Arum* inflorescence which was in the pollen-shedding or 'male' stage, they are likely to pollinate the pistillate flowers. It is only several hours later that the staminate or 'male' flowers dehisce and shower the prisoners with a rain of pollen. At that time, the barrier organs have wilted, the oil droplets and the stigmatic secretion are gone, as well as the heat and the smell, and the insects leave, covered with pollen which they may take to another *Arum* inflorescence which is still in the odoriferous, attractive 'female' stage.

For the plant physiologist, elucidation of the highly evolved timing mechanisms involved is a very appealing problem. *A priori* one can expect that these mechanisms are based on hormonal interactions between the various flower parts. *Sauromatum*, where male and female flowers are widely separated and where the time difference between the maturation of the female flowers and that of the male ones is more than 12 hours, is excellently suited for the study of interactions between floral parts.

Intensity and biological function of heat production, and the relationship between heat and smell

In many arum lilies, and in certain water lilies such as *Victoria* with a pollination system functionally similar to that of *Arum*, there can be no doubt that the heat produced by the inflorescence acts as a volatiliser for the smell (see Meeuse, 1975) and does not act as an attractant *per se*. Knoll demonstrated this convincingly in 1926 by showing that models of *Arum* inflorescences with small, heat-providing light-bulbs in the floral chamber failed to attract pollinators, whereas models provided with a mixture of rotting blood and glycerol (an anti-desiccant) did. In *Arum maculatum* the heat production can lead to a temperature difference with the environment of about 15 °C. In *Sauromatum* it is somewhat less, in *Schizocasia portei* larger (22 °C according to El-Din, 1968). Very high temperatures have been measured also in *Philodendron selloum* (where the heat is produced mostly by sterile male flowers: see Nagy, Odell & Seymour, 1972), and in a *Xanthosoma* species (B. J. D. Meeuse, unpublished). In *Arum dioscoridis* and *Sauromatum*, Meeuse (1973) has demonstrated the heat production visually by applying to the appendix, which in these forms is naturally dark-coloured, a thin film of a Vaseline-like mixture of liquid crystals; since the manner in which such crystals 'handle' light is temperature-dependent, a striking sequence of colours ranging from copper-red to peacock-blue manifested itself as the appendix, in the process of heating up, passed through the 25–28.5 °C range.

In most cases the substrate for the heat-producing respiration process is starch, but a lipid-fueled process appears to exist in *Schizocasia portei* (El-Din, 1968) and *Philodendron selloum* (Walker, 1980). In *Arum*, the dry weight of an (initially starch-rich) appendix may fall from about 32% to about 6% in the course of a single day. When respiration reaches its peak, the oxygen consumption may reach values of 72000 mm^3 O$_2$ g^{-1} wet wt h^{-1} (Lance, 1972), making the metabolism of the *Arum* appendix comparable to that of a hummingbird in flight. The metabolism of *Philodendron selloum* also compares favourably with that of hummingbirds and hawkmoths (Nagy *et al.*, 1972).

As already mentioned, heat production manifests itself also in the inflorescences of certain palms, cycads and *Pandanus utilis*. It should be interesting to investigate whether the phenomenon is correlated with odour production, and whether it is restricted to those palms that are insect-pollinated. The African oil-palm (*Elaeis guineensis*), whose inflorescences are heat-producing and display an anise-like odour, do appear to be beetle-pollinated.

In thermogenic water lily flowers such as those of *Nelumbo* and *Victoria* (for recent literature see Meeuse & Schneider, 1980) the pollination syndrome is similar to that of beetle-pollinated Araceae. Here too, then, heat acts as a volatiliser. In a broad sense, the production of odoriferous compounds such as amines and ammonia can, of course, be seen as an aspect of respiration (see Table 1), whether it be due to decarboxylation of amino acids as suggested by Simon (1962) and Richardson (1966) or to transamination of aldehydes, as championed by Hartmann, Ihlert & Steiner (1972) and Hartmann, Dönges & Steiner (1972).

Another function of thermogenicity in arum lilies may be that the high temperature it creates induces in the pollinating beetles (or other insects?) a readiness to mate. As a result, future generations of arum lilies would be assured of sufficient pollinators.

Finally, a third possible function of thermogenicity lies in the area of autecology. In eastern skunk cabbage (*Symplocarpus foetidus*) the developing inflorescences will, early in the year, when the ambient temperature may still be below freezing, push themselves up through the snow, which melts in their immediate vicinity. According to Knutson (1974, 1979) the plants possess a regulatory mechanism which guarantees relative temperature-constancy of the floral parts regardless of the prevailing environmental temperature: the colder it is, the more rapidly the inflorescences generate heat through respiration (Fig. 2). In this plant it is the bulky, starch-containing root which is said to keep providing fresh fuel. It would be interesting to investigate the transport capacity of this system. In Iowa, where Knutson carried out most of his investigations, self-pollination seems to be normal for *Symplocarpus*; later in the season, however, simulid flies and even honeybees may visit the inflorescences.

Table 1. *Differences in the 'spectrum' of odoriferous compounds produced by the inflorescence of various arum lilies*

	Ammonia	Indole	Skatole	Trimethyl-amine	Other amines
Arum dioscoridis	+	(+)	+ + +	−	4
Arum italicum	+	−	+	−	11
Dracunculus	+	+ +	−	+ + +	4
Sauromatum	+	+ + +	−	+	13

+, present; −, absent.

Fig. 2. Maintenance of temperature stability by inflorescences of Eastern skunk cabbage (*Symplocarpus foetidus*). The lower the ambient temperature the higher the rate of respiration.

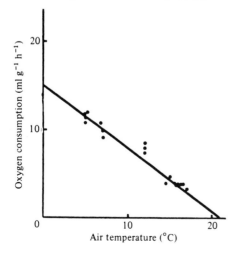

The biochemical basis for thermogenicity in arum lilies, and the triggering of the metabolic explosion in the appendix

There is now a consensus that thermogenicity in plant tissues is based on the presence of a dual pathway for respiratory electron transport: the classical, cyanide-sensitive chain which is coupled to the generation of high-energy phosphate (ATP), and a cyanide-resistant alternative pathway which can be inhibited by propyl gallate and derivatives of hydroxamic acid. The cyanide-resistant pathway is phosphorylative to a much lesser extent, and perhaps not at all in special cases or under special circumstances (Passam & Palmer, 1972).

When electrons coming from the respiratory substrate are forced to go through the alternative pathway (e.g. by the addition of cyanide), the energy available from oxidation of the substrate is no longer 'trapped' in the form of ATP but appears immediately as heat. To what extent, and when, the alternative pathway is operative under natural conditions (i.e. in the absence of cyanide) is still a matter of dispute, in spite of an impressive amount of work done in recent years on cyanide-resistant respiration and the factors that regulate or control it (see Henry & Nyns, 1975; Solomos, 1977; Day *et al.*, 1980; Storey, 1980; Laties, 1982). Methods for assessing the relative contribution of the cyanide-insensitive pathway to the overall respiration in plant tissues have been worked out (Theologis & Laties, 1978a, b; Lambers & Smakman, 1978). A good case can be made for the idea that membrane integrity, which requires the synthesis of certain lipids and proteins, is essential for the operation of the cyanide-insensitive pathway. Restoration of cyanide-insensitivity in (e.g.) potato slices upon ageing could probably be explained on this basis (Day *et al.*, 1980).

The importance of ascorbic acid, demonstrated by Arrigoni, Arrigoni-Liso & Calabrese (1976) and Arrigoni, De Santis & Calabrese (1977), is probably due to the fact that this compound is required for the synthesis of a hydroxyproline-rich protein which in turn may play a role in cyanide-resistant mitochondria.

Although wasteful in a biochemical sense, cyanide-insensitive respiration may provide the plant cell with a means of obtaining building-blocks for cellular metabolism without being swamped with ATP (Palmer, 1976). In those cases where there is marked thermogenicity, such as in arum lilies, operation of the alternative pathway has survival value (also) in an ecological sense. 'No heat' (that is, exclusive operation of the classical pathway) would mean no volatilisation of the odoriferous compounds that attract the pollinators, and consequently there would be no insect visitors and no outbreeding with its built-in genetic and evolutionary advantages.

Cyanide-insensitive respiration was discovered independently by Okunuki in pollen and by van Herk in the *Sauromatum* appendix (see Meeuse, 1975, for references). The latter investigator has also succeeded in demonstrating that the metabolic explosion in the *Sauromatum* appendix is unleashed by a hormone (calorigen) which originates in the primordia (buds) of the staminate flowers (see Fig. 1) and begins to leave these to move into the appendix about 22 hours before the metabolic peak is reached. Having developed a bioassay based on the formation of indole by pieces of immature appendix treated with calorigen-containing extracts, Meeuse and co-workers have succeeded in isolating and purifying two compounds with calorigen activity (Buggeln & Meeuse, 1971; Chen & Meeuse, 1975). These are fairly stable, low-molecular-weight compounds (less than 1000 daltons), probably possessing a carboxyl group. Data have been obtained on their ultraviolet and infra-red absorption, fluorescence and nuclear magnetic resonance spectra. Difficulties were encountered in obtaining definitive mass spectroscope data, due to decomposition of the calorigen.

The long lag time between the moment of calorigen application and the appearance of the metabolic peak argues in favour of the idea that calorigen-triggered synthesis of new enzymatic protein is essential. Using cycloheximide and actinomycin A, McIntosh & Meeuse (1978) have obtained some evidence in favour of this.

The role of the light/dark regime in the anthesis of *Sauromatum* and *Arum*

Confirming earlier observations on *Arum* by Schmucker (1925), Buggeln, Meeuse & Klima (1971) (see also Meeuse & Buggeln, 1969) demonstrated that normal anthesis in *Sauromatum* requires an alternation of light and dark periods. When *Sauromatum* inflorescences raised in constant light were given a 'dark shot' of at least six hours at the time they would normally have opened (or even 1–2 days later), anthesis followed, with the metabolic peak occurring 42–45 hours (i.e. almost two days) after the beginning of the dark period (Fig. 3). A literature search revealed that the long lag time (which initially appeared surprising) is not unique in flower opening. For instance, in experiments on the flowers of evening primroses (*Oenothera*) Arnold (1959) could show that after a reversal of day and night was instituted, the mature buds continued for two days to open (synchronously) at their normal time of 6 p.m.; it was only after two full days that the buds switched to the 'new' opening time of 6 a.m. This observation is far from trivial, since it demonstrates that the unfolding of a flower or inflorescence, rather than being a 'simple' event based (e.g.) on quick turgor changes in petal-cells, is a carefully 'planned', programmed phenomenon; two days before their expected opening time the *Oenothera* buds are already committed to a certain course of action, and a change in regime imposed after that point (i.e. within the two-day time-span) has little or no effect.

Possible participation of ethylene

That ethylene is somehow involved in the flowering events of arum lilies is *a priori* likely when one considers the role which the gas demonstrably plays in post-pollination phenomena, senescence, and certain timing events in flowers resulting in dichogamy (Cooke & Randall, 1968; Marei & Crane, 1971; Zeroni, Ben Yehoshua & Galil, 1972; Galil, Zeroni & Bar Shalom, 1973; Blumenfeld, 1975; Lieberman, 1979; Rhodes, 1980). Further indirect evidence has been provided by workers of the Laties school, who have claimed a link between ethylene and cyanide-insensitive respiration. The reactions of certain types of fruits and of potatoes to either ethylene or cyanide are very similar (Solomos & Laties, 1973; Goldmann & Laties, 1976; Solomos, 1977; Day, Arron, Christoffersen & Laties, 1978; see also Reid & Pratt, 1972, and Rychter, Janes & Frenkel, 1978, 1979). Either agent leads to a sharp rise in overall respiration, concomitant with a sharp increase in the concentration of fructose bisphosphate and a decline in glucose-6-phosphate and phosphoenolpyruvate (Solomos & Laties, 1974). In the climacteric of fruit, an increase in ATP has also been observed in some cases (Solomos & Laties, 1976a, b). This could be explained by an increased electron flow through substrate-level phosphorylation sites and through site I of oxidative phosphorylation. The changes in glycolytic intermediates and in ATP observed in the *Sauromatum* appendix during the natural flowering sequence are in agreement with the above data (Hess & Meeuse, 1968a, b; Meeuse, Buggeln & Summers, 1969). The viewpoint that the ethylene-induced surge in

Fig. 3. Methods used to compute the time budgets for the induction of anthesis in *Sauromatum*. □, amputation of appendix; △, metabolic peak.

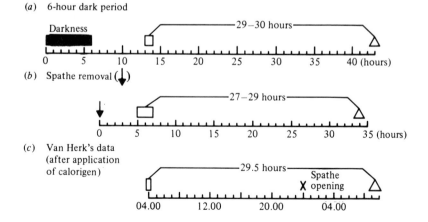

(a) 6-hour dark period

Darkness   29–30 hours

0   5   10   15   20   25   30   35   40 (hours)

(b) Spathe removal (↓)

27–29 hours

0   5   10   15   20   25   30   35 (hours)

(c) Van Herk's data (after application of calorigen)

29.5 hours   Spathe ✗ opening

04.00   12.00   20.00   04.00

respiration is due to an induction of the alternative pathway could not be maintained in view of the observation that the respiratory surge and the increase in glycolysis would manifest themselves even in the presence of *m*-chlorobenzhydroxamic acid, a potent inhibitor of the cyanide-insensitive alternative pathway. It thus seems more likely that ethylene induces an increase in glycolytic substrates and that this is brought about by a spectacular increase in phosphofructokinase and pyruvate kinase.

Meeuse & Buggeln (1969) and Buggeln (1969) have provided evidence of a somewhat more direct nature when they demonstrated that removal of the spathe at least two days before natural blooming time has the same effect as the light/dark treatment that induces the metabolic explosion in the appendix. Analysing the time budgets for the various modes of induction (by calorigen, by spathe removal and by light/dark treatment), Buggeln has convincingly argued that spathe removal and calorigen application act on one and the same process (see Fig. 3). Making razor-slashes in the floral chamber has the same effect as spathe extirpation, and this makes it likely that wound-ethylene is involved rather than the removal of an inhibitor one could postulate to be present in the spathe.

References

Arnold, C.-G. (1959). Die Blütenöffnung bei *Oenothera* im Abhängigkeit vom Licht-Dunkel-Rhythmus. *Planta Berlin*, **53**, 198–211.

Arrigoni, O., Arrigoni-Liso, R. & Calabrese, G. (1976). Ascorbic acid as a factor controlling the development of cyanide-insensitive respiration. *Science*, **194**, 332–3.

Arrigoni, O., De Santis, A. & Calabrese, G. (1977). The increase of hydroxyproline-containing proteins in Jerusalem artichoke mitochondria during the development of cyanide-insensitive respiration. *Biochemical and Biophysical Research Communications*, **74**, 1637–40.

Blumenfeld, A. (1975). Ethylene and the *Annona* flower. *Plant Physiology*, **55**, 265–9.

Buggeln, R. G. (1969). Control of blooming in *Sauromatum guttatum* Schott (Araceae). PhD thesis, University of Washington, Seattle.

Buggeln, R. G. & Meeuse, B. J. D. (1971). Hormonal control of the respiratory climacteric in *Sauromatum guttatum* (Araceae). *Canadian Journal of Botany*, **49**, 1373–7.

Buggeln, R. G., Meeuse, B. J. D. & Klima, J. R. (1971). Control of blooming in *Sauromatum guttatum* Schott (Araceae) by darkness. *Canadian Journal of Botany*, **49**, 1025–31.

Chen, J. & Meeuse, B. J. D. (1975). Purification and partial characterization of two biologically active compounds from the inflorescence of *Sauromatum guttatum* Schott. *Plant and Cell Physiology*, **16**, 1–11.

Cooke, A. R. & Randall, I. D. (1968). 2-Haloethane phosphonic acids as ethylene-releasing reagents for the induction of flowering in pine-apples. *Nature, London*, **281**, 974.

Day, D. A., Arron, G. P., Christoffersen, R. E. & Laties, G. G. (1978). Effect of ethylene and carbon dioxide on potato metabolism. *Plant Physiology*, **62**, 820–5.

Day, D. A., Arron, G. P. & Laties, G. G. (1980). Nature and control of respiratory pathways in plants: the interaction of cyanide-resistant respiration with the cyanide-sensitive pathway. In *The Biochemistry of Plants*, vol. 2, ed. P. K. Stumpf & E. E. Conn, pp. 198–241. New York & London: Academic Press.

El-Din, S. M. (1968). Wärmeperiode und Duftstoff des Blütenkolbens der Araceae *Schizocasia portei*. *Naturwissenschaften*, **12**, 658–9.

Galil, J., Zeroni, M. & Bar Shalom (Bogoslavsky), D. (1973). Carbon dioxide and ethylene effects in the coordination between the pollinator *Blastophaga quadraticeps* and the syconium in *Ficus religiosa*. *New Phytologist*, **72**, 1113–27.

Goldman, D. & Laties, G. G. (1976). The initiation of the climacteric rise in peeled banana fruit by HCN and CO under hypobaric conditions. *Plant Physiology (Supplement)*, **57**, 27.

Hartmann, T., Dönges, D. & Steiner, M. (1972). Biosynthese aliphatischer Monoamine in *Mercurialis perennis* durch Aminosäure–Aldehyd-Transaminierung. *Zeitschrift für Pflanzenphysiologie*, **67**, 404–17.

Hartmann, T., Ihlert, H. I. & Steiner, M. (1972). Aldehydaminierung, der bevorzügte Biosyntheseweg für primäre, aliphatische Mono-amine in Blütenpflanzen. *Zeitschrift für Pflanzenphysiologie*, **68**, 11–18.

Henry, M. F. & Nyns, E. J. (1975). Cyanide-resistant respiration. An alternate mitochondrial pathway. *Sub-Cellular Biochemistry*, **4**, 1–65.

Hess, C. M. & Meeuse, B. J. D. (1968a). Factors contributing to the respiratory flare-up in the appendix of *Sauromatum* (Araceae): I. *Verhandelingen van de Koninklyke Nederlandse Akademie van Wetenschappen, Amsterdam, Serie C*, **74**, 443–55.

Hess, C. M. & Meeuse, B. J. D. (1968b). Factors contributing to the respiratory flare-up in the appendix of *Sauromatum* (Araceae). II. *Verhandelingen van de Koninklyke Nederlandse Akademie van Wetenschappen, Amsterdam, Serie C*, **74**, 456–71.

Knoll, F. (1926). Insekten und Blumen. Experimentelle Arbeiten zur Vertiefung unserer Kenntnisse über die Wechselbeziehungen zwischen Pflanzen und Tieren. IV. Die *Arum*-Blütenstände und ihre Besucher. *Abhandlungen der Zoologisch-Botanischen Gesellschaft in Wien*, **12**, 279–482.

Knutson, R. M. (1974). Heat-production and temperature regulation in Eastern Skunk Cabbage. *Science*, **186**, 746–7.

Knutson, R. M. (1979). Plants in heat. *Natural History*, **88**, 42–7.

Lambers, H. & Smakman, G. (1978). Respiration of flood-tolerant and flood-intolerant *Senecio* species. Affinity for oxygen and resistance to cyanide. *Physiologia Plantarum*, **42**, 163–6.

Lance, C. (1972). La respiration de l'*Arum maculatum* au cours du développement de l'inflorescence. *Annales des sciences naturelles, botanique, 12th Series*, **13**, 477–95.

Laties, G. G. (1982). The cyanide-resistant, alternative path in higher plant respiration. *Annual Review of Plant Physiology*, **33**, 519–55.

Lieberman, M. (1979). Biosynthesis and action of ethylene. *Annual Review of Plant Physiology*, **30**, 533–91.

McIntosh, L. & Meeuse, B. J. D. (1978). Control of the development of cyanide-resistant respiration in *Sauromatum guttatum* (Araceae). In *Plant Mitochondria*, ed. G. Ducet & C. Lance, pp. 339–45. Amsterdam: Elsevier/North-Holland Biomedical Press.

Marei, N. & Crane, C. J. (1971). Growth and respiratory response of fig (*Ficus carica* L. cv. Mission) fruits to ethylene. *Plant Physiology*, **48**, 240–54.

Meeuse, B. J. D. (1973). Films of liquid crystals as an aid in pollination-studies. In *Pollination and Dispersal*, ed. N. B. M. Brantjes, pp. 19–20. Nymegen: University of Nymegen.

Meeuse, B. J. D. (1975). Thermogenic respiration in aroids. *Annual Review of Plant Physiology*, **26**, 117–26.

Meeuse, B. J. D. & Buggeln, R. G. (1969). Time, space, light and darkness in the metabolic flare-up of the *Sauromatum* appendix. *Acta Botanica Neerlandica*, **51**, 159–72.

Meeuse, B. J. D., Buggeln, R. G. & Summers, S. N. (1969). Nucleotide levels in the appendix of *Sauromatum* and *Arum* (Araceae) during the flowering sequence. In *Abstracts of the 9th International Botanical Congress, Seattle 1969*, p. 144.

Meeuse, B. J. D. & Schneider, E. (1980). *Nymphaea* revisited – a preliminary communication. *Israel Journal of Botany*, **28**, 65–79.

Nagy, K. A., Odell, D. K. & Seymour, R. S. (1972). Temperature regulation by the inflorescence of *Philodendron*. *Science*, **178**, 1195–7.

Palmer, J. M. (1976). The organization and regulation of electron transport in plant mitochondria. *Annual Review of Plant Physiology*, **27**, 113–57.

Passam, H. C. & Palmer, J. M. (1972). Electron transport and oxidative phosphorylation in *Arum* spadix mitochondria. *Journal of Experimental Botany*, **23**, 366–74.

Reid, M. S. & Pratt, H. K. (1972). Effects of ethylene on potato tuber respiration. *Plant Physiology*, **49**, 252–5.

Rhodes, M. H. C. (1980). Respiration and senescence of plant organs. In *The Biochemistry of Plants*, vol. 2, ed. P. K. Stumpf & E. E. Conn, pp. 419–62. New York & London: Academic Press.

Richardson, I. (1966). Studies on the biogenesis of some simple amines and quaternary ammonium compounds in higher plants. Isoamylamine and isobutylamine. *Phytochemistry*, **5**, 23–30.

Rychter, A., Janes, H. W. & Frenkel, C. (1978). Cyanide-resistant respiration in freshly cut potato slices. *Plant Physiology*, **61**, 667–8.

Rychter, A., Janes, H. W. & Frenkel, C. (1979). Effect of ethylene and oxygen on the development of cyanide-resistant respiration in white potato mitochondria. *Plant Physiology*, **63**, 149–51.

Schmucker, T. (1925). Beiträge zur Biologie und Physiologie von *Arum maculatum*. *Flora*, **118**, 460–75.

Simon, E. W. (1962). Valine decarboxylation in *Arum* spadix. *Journal of Experimental Botany*, **13**, 1–4.

Solomos, T. (1977). Cyanide-resistant respiration in higher plants. *Annual Review of Plant Physiology*, **28**, 279–97.

Solomos, T. & Laties, G. G. (1973). Cellular organization and fruit ripening. *Nature, London*, **245**, 390–2.

Solomos, T. & Laties, G. G. (1974). Similarities between the actions of ethylene and cyanide in initiating the climacteric and ripening of avocados. *Plant Physiology*, **54**, 506–11.

Solomos, T. & Laties, G. G. (1976a). Induction by ethylene of cyanide-resistant respiration. *Biochemical and Biophysical Research Communications*, **70**, 663–71.

Solomos, T. & Laties, G. G. (1976b). Effects of cyanide and ethylene on the respiration of cyanide-sensitive and cyanide-resistant plant tissues. *Plant Physiology*, **58**, 47–50.

Storey, B. T. (1980). Electron transport and energy coupling in plant mitochondria. In *The Biochemistry of Plants*, vol. 2, ed. P. K. Stumpf & E. E. Conn, pp. 125–95. New York & London: Academic Press.

Theologis, A. & Laties, G. G. (1978a). Relative contribution of cytochrome-mediated and cyanide-resistant electron transport in fresh and aged potato slices. *Plant Physiology*, **62**, 232–7.

Theologis, A. & Laties, G. G. (1978b). Respiratory contribution of the alternate path during various stages of ripening of avocado and banana fruits. *Plant Physiology*, **62**, 249–55.

Vogel, S. (1963). Duftdrüsen im Dienste der Bestäubung. *Abhandlungen der Mathematisch-Naturwissenschaftlichen Klasse, Akademie der Wissenschaften und der Literatur, Mainz*, **10**, 599–763.

Walker, D. B. (1980). Structural and histochemical study of the heat-generating, sterile, male flowers in *Philodendron selloum*. In *Botany 80 UBC (Abstracts of papers presented at the University of British Columbia, Vancouver, 12–16 July 1980)*, Botanical Society of America, Miscellaneous Series, Publication 158, p. 122.

Zeroni, M., Ben Yehoshua, S. & Galil, J. (1972). Relationship between ethylene and the growth of *Ficus sycomorus*. *Plant Physiology*, **50**, 378–81.

LINUS H. W. van der PLAS

5 Respiration and morphogenesis in storage tissue

Storage tissues are widely used in research on plant respiration. In particular the respiratory rise seen during the first days after wounding the tissue (induced respiration) has been extensively studied (see reviews of Kahl, 1973, 1974; Van Steveninck, 1975; Laties, 1978). This increase in respiration develops during the dedifferentiation of the cells beneath the wound surface and depends on the synthesis of RNA and protein. This dedifferentiation is followed by a redifferentiation phase (suberisation, periderm formation, callus outgrowth). The morphogenetic response varies with the plant species, but it also depends on tissue age and on the incubation conditions. Even in tissue of one species a wide range of morphogenetic responses is possible; this is summarised for potato tissue in Fig. 1.

Respiratory characteristics of storage tissue during different morphogenetic responses
Quantitative and qualitative differences
  Differences in respiratory properties with different morphogenetic responses of storage tissue may be both quantitative and qualitative.

Fig. 1. Various morphogenetic responses of potato tuber tissues upon slicing. Data derived from Akemine, Kikuta & Tagawa (1975), Kahl, Rosenstock & Lange (1969), Kahl (1973), Lange & Rosenstock (1965) and Lange (1970).

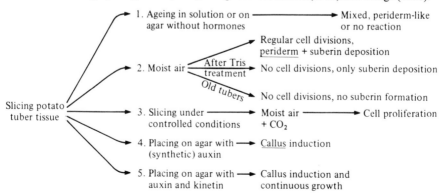

When periderm is formed the induced respiratory rise is generally transient (Fig. 2). After the 'closing of the wound' the cells beneath the wound surface become metabolically less active again and the level of respiration returns to that of the intact organ. During callus induction and growth the cells remain metabolically active as long as cell divisions go on and respiration maintains a high level. Overall respiratory activity reflects the need for energy and intermediates during redifferentiation.

Qualitative differences in respiration during the various morphogenetic responses can arise, when a branched system of sugar degradation and electron transport offers the cells the opportunity to make a choice between 'concurrent' pathways. Important branch-points may be (Fig. 3): the choice between the pentose phosphate (PP) pathway and glycolysis at the level of

Fig. 2. Respiratory activity of potato tuber tissue after wounding and during differentiation in different directions. (Data derived from Lange, 1970; Rosenstock, Kahl & Lange, 1971; L. H. W. van der Plas, unpublished.)

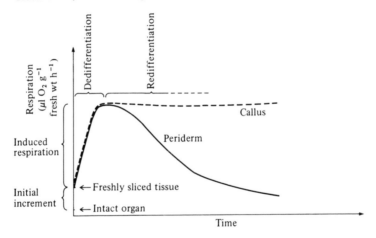

Fig. 3. Schematic representation of sugar degradation and electron transport routes in plants. CoQ, coenzyme Q; G-6-P, glucose-6-phosphate; PP, pentose phosphate; Pyr, pyruvate.

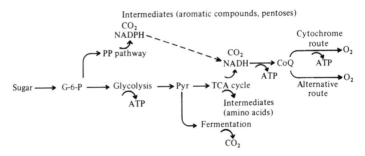

glucose-6-phosphate; the choice between the tricarboxylic acid (TCA) cycle and fermentation to the level of pyruvate; and the choice between the cytochrome and the cyanide-resistant alternative electron transport route at the level of coenzyme Q. In addition the choice between various (mitochondrial) NADH dehydrogenases may be important.

A change in the usage of particular pathways causes differences in the amounts of ATP, NADH/NADPH and intermediates produced and results in adaptation of the production of these essential compounds to the needs of the cells. The regulation of this adaptation may take place at the level of the capacity and/or the activity of the pathways.

Differences in capacity

An increase in the capacity of the sugar degradation system is generally found after wounding of storage tissue. However, the increase in capacity of the individual pathways varies and may be different in the various morphogenetic responses. This leads to differences in the relative capacities of these pathways following morphogenesis. Apparently, the cells have the potential to manipulate the induction of the various pathways of the system separately. The regulation of this induction is influenced by various factors. Table 1 summarises for potato some changes observed in endogenous hormone concentrations during storage and wounding and some effects of application of these hormones on morphogenesis and respiration. Although the available information is often scarce and incomplete, the importance of plant hormones in regulating morphogenesis and respiratory characteristics is obvious. Yet there are other factors which may cause changes in the relative capacities of the various pathways of the sugar degradation system. In potato the presence of substances such as ethanol, chloramphenicol or sucrose in the callus nutrient medium and changes in growth temperature affect the relative alternative oxidase capacity (van der Plas & Wagner, 1980*b* and unpublished).

Differences in activity

Morphogenesis-dependent differences in the activity of 'competitive' branches are described for the PP pathway versus glycolysis/TCA cycle. Some results for potato tissue are summarised in Fig. 4. Conclusions regarding the relative activity of these pathways are generally drawn from experiments using malonate as a (specific TCA cycle) inhibitor, from experiments on the release of $1\text{-}^{14}CO_2$ and $6\text{-}^{14}CO_2$ from ^{14}C-labelled glucose, and from determinations of the levels of NAD(P)(H). It appears that during the lag phase of callus induction the PP pathway becomes relatively more important, which is ascribed to the greater need for NADPH and intermediates of the PP pathway

Table 1. *Changes in (endogenous) plant hormone levels during storage and wounding in potato; and some effects of plant hormones on morphogenesis and respiration in potato*

	Changes in hormone levels		Effect of applied hormone	
	With storage	After wounding	On morphogenesis	On respiration
Indoleacetic acid	? (Increase with sprouting)	Increase	Stimulates periderm formation. High concentrations: some callus formation	Relatively greater capacity of alternative oxidase
Synthetic auxins	—	—	Low concentrations stimulate periderm formation, high concentrations callus formation	Greater activity of PP pathway in lag phase callus induction. Relatively greater capacity alternative oxidase
Cytokinins	?	Increase	Stimulates continuous callus growth. Inhibits suberisation	Greater activity of PP pathway during callus growth
Gibberellins	Increase	?	Little or no effect	Some overall stimulation or no effect
Abscisic acid	Decrease	Increase	Stimulates suberisation. Inhibits callus formation	Some overall stimulation. Relatively smaller capacity of alternative oxidase
Ethylene	?	Transient increase	Little effect?	Greater capacity of alternative oxidase

Data derived from: Akemine, Kikuta & Tagawa (1970, 1975), Bialek & Bielinska-Czarnecka (1975), Conrad & Köhn (1975), Coutrez-Geerinck (1973), Hourmant & Penot (1978), Kikuta, Akemine & Tagawa (1971), Kikuta et al. (1977), Laties (1978), Lindblom (1968), McGlasson (1969), Rosenstock & Kahl (1978), Rychter, Janes & Frenkel (1979), Soliday, Dean & Kolattukudy (1978), van der Plas & Wagner (1980a, b and unpublished).

General remarks: High concentrations of the various hormones generally inhibit respiration. Hormone combinations may give aberrant results. Some of the hormone effects may be indirect, since some hormones influence the production of other hormones.

for various biosynthetic reactions during subsequent callus growth. In the presence of both auxin and cytokinin cell division in the callus goes on, the extra need for NADPH and intermediates persists and the relative activity of the PP pathway remains high (Fig. 4, line 5). In the presence of only auxin an initial stage with cell divisions is followed by a stage with predominant cell expansion during which the relative activity of the PP pathway decreases (Fig. 4, line 4). Considering these and other changes in relative activities, it should be kept in mind that the total respiration may vary (Fig. 2): so a decrease in the *relative* activity of glycolysis/TCA cycle (during the first days after wounding) may still be accompanied by an increase in the *absolute* activity. However, the increase in PP pathway activity is greater then.

Also, a direct relation between suberisation and TCA cycle activity has been described (Lange, 1970). Cytokinins which stimulate callus growth and PP pathway activity inhibit this suberisation (Soliday, Dean & Kolattukudy, 1978).

As regards the other 'competitive' pathways (Fig. 3), it is doubtful whether the alternative cyanide-resistant electron transport route is used *in vivo* in

Fig. 4. Changes in the relative activities of the PP pathway and glycolysis/TCA cycle during periderm or callus induction in potato. 1, Tissue placed on agar without hormones; 2, tissue placed in moist air; 3, proliferating tissue (slices in carbon dioxide); 4, tissue placed on agar with a (synthetic) auxin; 5, tissue placed on agar with auxin and cytokinin. High = high inhibition of respiration by malonate (a TCA cycle inhibitor); high ratio of $6\text{-}^{14}CO_2$ to $1\text{-}^{14}CO_2$ released from ^{14}C-labelled glucose; high ratio of glycolysis/TCA cycle to PP pathway. Low = low values of these parameters. (Adapted from Akemine, Kikuta & Tagawa, 1970, 1975; Kikuta, Akemine & Tagawa, 1971; Kikuta *et al.*, 1977; Lange, 1970.)

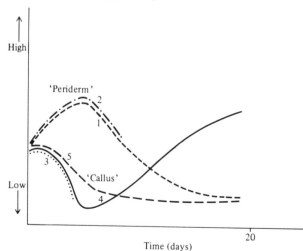

storage tissue slices (Laties, 1978). However, under some circumstances (low temperatures) alternative oxidase may function *in vivo* at least in callus from some plant species (Yoshida & Tagawa, 1979). Functioning of (alcoholic) fermentation has been described during aerobic incubation of callus-forming carrot tissue (Komamine, Morohashi & Shimokoriyama, 1969), but does not take place in callus-forming potato tuber tissue (L. H. W. van der Plas, unpublished).

Apparently, during morphogenesis various possibilities are available to the cell to direct degradation of the carbon source and electron transport along various routes, yielding various products in different amounts. The question remains at the moment whether and when the cell uses these possibilities and how this process is regulated.

References

Akemine, T., Kikuta, Y. & Tagawa, T. (1970). Respiratory changes during callus formation in potato tuber tissue cultured *in vitro. Journal of the Faculty of Agriculture, Hokkaido University, Sapporo*, **56**, 323–36.

Akemine, T., Kikuta, Y. & Tagawa, T. (1975). Effect of kinetin and naphthalene-acetic acid application on the respiratory metabolism during callus development in potato tuber tissue cultured *in vitro. Journal of the Faculty of Agriculture, Hokkaido University, Sapporo*, **58**, 247–61.

Bialek, K. & Bielinska-Czarnecka, M. (1975). Gibberellinlike substances in potato tubers during their growth and dormancy. *Bulletin de l'Académie poloniaise des sciences*, **23**, 213–18.

Conrad, K. & Köhn, B. (1975). Zunahme von Cytokinin und Auxin in verwundeten Speichergewebe von *Solanum tuberosum. Phytochemistry*, **14**, 325–8.

Coutrez-Geerinck, D. (1973). Consommation d'oxygène modifiée par l'acide gibberellique, chez des tissus fragmentés de tubercules de *Solanum tuberosum* en fonction de l'âge de ces organes. *Bulletin de l'Académie royale de belgique*, **59**, 948–55.

Hourmant, A. & Penot, M. (1978). Action de quelques phytohormones sur la respiration et l'absorption du phosphate par des disques de tubercules de pomme de terre en survie. *Physiologia Plantarum*, **44**, 278–82.

Kahl, G. (1973). Genetic and metabolic regulation in differentiating plant storage tissue cells. *Botanical Review*, **39**, 274–99.

Kahl, G. (1974). Metabolism in plant storage tissue slices. *Botanical Review*, **40**, 263–314.

Kahl, G., Rosenstock, G. & Lange, H. (1969). Die Trennung von Zellteilung und Suberinsynthese in dereprimiertem pflanzlichem Speichergewebe durch Tris-(hydroxymethyl)-aminomethan. *Planta*, **87**, 365–71.

Kikuta, Y., Akemine, T. & Tagawa, T. (1971). Role of pentose phosphate pathway during callus development in explants from potato tuber. *Plant and Cell Physiology*, **12**, 73–9.

Kikuta, Y., Harada, T., Akemine, T. & Tagawa, T. (1977). Role of kinetin in activity of the pentose phosphate pathway in relation to growth of potato tissue cultures. *Plant and Cell Physiology*, **18**, 361–70.

Komamine, A., Morohashi, Y. & Shimokoriyama, M. (1969). Changes in respiratory metabolism in tissue cultures of carrot root. *Plant and Cell Physiology*, **10**, 411–24.

Lange, H. (1970). Atmungswege bei vernarbenden und proliferierenden Gewebefragmenten der Kartoffelknolle. *Planta*, **90**, 119–32.

Lange, H. & Rosenstock, G. (1965). Kausanalytische Untersuchungen zum Alterungsvorgang bei der Wundkompensation von Speicherorganen unter besonderer Berücksichtigung des Nährstoffaktors. *Phytopathologische Zeitschrift*, **52**, 188–201.

Laties, G. G. (1978). The development and control of respiratory pathways in slices of plant storage organs. In *Biochemistry of Wounded Plant Tissues*, ed. G. Kahl, pp. 421–66. Berlin & New York: Walter de Gruyter.

Lindblom, H. (1968). Sprouting behaviour in relation to storage conditions and to indole-3-acetic acid, gibberellins and inhibitor β in seed potato tubers. *Acta Agriculturae Scandinavica*, **18**, 177–95.

McGlasson, W. B. (1969). Ethylene production by slices of green banana fruit and potato tuber tissue during the development of induced respiration. *Australian Journal of Biological Sciences*, **22**, 489–91.

Rosenstock, G. & Kahl, G. (1978). Phytohormones and the regulation of cellular processes in aging storage tissues. In *Biochemistry of Wounded Plant Tissues*, ed. G. Kahl, pp. 623–71. Berlin & New York: Walter de Gruyter.

Rosenstock, G., Kahl, G. & Lange, H. (1971). Beziehungen zwischen Entwicklungszustand und Wundatmung beim Speicherparenchym von *Solanum tuberosum*. *Zeitschrift für Pflanzenphysiologie*, **64**, 130–8.

Rychter, A., Janes, H. W. & Frenkel, C. (1979). Effect of ethylene and oxygen on the development of cyanide-resistant respiration in whole potato mitochondria. *Plant Physiology*, **63**, 149–51.

Soliday, C. L., Dean, B. B. & Kolattukudy, P. E. (1978). Suberization: inhibition by washing and stimulation by abscisic acid in potato disks and tissue culture. *Plant Physiology*, **61**, 170–4.

van der Plas, L. H. W. & Wagner, M. J. (1980*a*). Changes in alternative oxidase activity and other mitochondrial parameters with callus formation by potato tuber tissue discs. *Plant Science Letters*, **17**, 207–13.

van der Plas, L. H. W. & Wagner, M. J. (1980*b*). Influence of ethanol on alternative oxidase in mitochondria from callus-forming potato tuber discs. *Physiologia Plantarum*, **49**, 121–6.

Van Steveninck, R. F. M. (1975). The 'washing' or 'aging' phenomenon in plant tissues. *Annual Review of Plant Physiology*, **26**, 237–58.

Yoshida, S. & Tagawa, F. (1979). Alteration of the respiratory function in chill-sensitive callus due to low temperature stress. I. Involvement of the alternative pathway. *Plant and Cell Physiology*, **20**, 1243–50.

R. M. M. CRAWFORD

6 Anaerobic respiration and flood tolerance in higher plants

Natural anaerobiosis

Completely anaerobic conditions are probably rare in plant tissues. Nevertheless, at certain times in germinating seeds (Leblova, 1978) and in flooded roots and rhizomes (Öpik, 1980) the rate of oxygen supply is not sufficient to suppress the accumulation of the products of glycolysis. The minimal oxygen concentration which is necessary to prevent anaerobic fermentation exceeding aerobic metabolism is termed the extinction point (Blackman, 1928). It is not constant for a given organ but depends on physiological state, temperature and the diffusive resistance to oxygen within the tissue (Turner, 1951). Below the extinction point and the final disappearance of oxygen the tissues are in a state of hypoxia. The readiest example of a hypoxic state in higher plants is in germinating seeds. Before the rupture of the testa and the emergence of the seedling most seeds pass through a state in which there is a temporary accumulation of glycolytic end-products (Sherwin & Simon, 1969; Leblova, 1978). Under normal conditions in well-aerated soils this period of partial anaerobiosis rarely lasts longer than 48 hours. However, flooding of the soil can prolong this condition. While this will damage some species, others are able to germinate even after prolonged periods of anaerobiosis (Crawford, 1978).

In many herbaceous perennials of wetland sites the perennating organ, be it root or rhizome, can be completely buried beneath the surface of an anaerobic soil. If the buried organ has even a dead stalk persisting above the level of the flood-water then this can provide a channel for the downward diffusion of oxygen to the submerged rhizome (Brändle, 1980). Nevertheless, buried rhizomes of some species can survive even when there is no aerial supply of oxygen. The bulrush (*Schoenoplectus lacustris*) is one such species that can send up new shoots even when the rhizome is completely buried.

During the growing season the plants of wetland sites have an amphibious existence. Flooding may submerge the roots in an anaerobic environment but the shoots usually have access to free air which diffuses internally to the roots (Armstrong, 1980). Whether or not this downward diffusion of oxygen is sufficient to keep the active regions of the root above the respiratory compensation point is still the subject of debate (Hook & Crawford, 1978). When roots of rice and maize are exposed to an atmosphere of nitrogen with their shoots in air, there is an immediate fall in root ATP relative to ADP and AMP, i.e. energy charge (Raymond, Bruzau & Pradet, 1978), indicating that internal diffusion cannot supply all the oxygen needs of the root. In tree species tolerant of flooding, oxygen diffuses internally to the submerged roots yet at the same time ethanol accumulates (Hook & Brown, 1973; Crawford & Baines, 1977). These and other examples show that both aerobic and anaerobic metabolism contribute to the respiratory needs of flooded roots. It is therefore not clear whether flood-tolerant plants survive solely because they receive an internal supply of oxygen or whether anaerobic activity helps to overcome periods of oxygen stress.

Endurance under total anoxia

The amphibious nature of flood-tolerant species noted above, makes it difficult to judge in intact plants the importance of an anaerobic metabolism. Experimentally, various plant organs can be tested for their ability to survive periods of oxygen starvation (anoxia) and thus show their probable response to this stress in the field. Intact organs can be incubated for prolonged periods in anaerobic jars and incubators where residual traces of oxygen are removed by circulation with hydrogen over a palladium catalyst. If these organs possess non-dormant buds the effect of the period of oxygen starvation can be examined on their subsequent ability to resume growth in air. Table 1 lists a number of wetland species which have been tested for their capacity to resume growth in air after having been kept in an anaerobic incubator for seven days at 22 °C. The experiment allows this group of flood-tolerant plants to be divided into two main classes, namely those that can and those that cannot survive prolonged anoxia. It should be noted that *Glyceria maxima*, a species which has been frequently studied in relation to its capacity to grow in wet soils (Chirkova, 1978; Crawford, 1978; Smith & ap Rees, 1979), is killed within 3–4 days of total anoxia, as is rice and all the species of *Juncus* so far tested.

The most remarkable observation from this study is that among the species tolerant of anoxia there are a number which are capable of sustained production of new healthy shoots during the period of anoxia (Figs. 1 and 2). It is sometimes claimed that the rice coleoptile is the only plant organ able to grow in a complete absence of oxygen (Pradet & Bomsel, 1978). The ability of rice to tolerate a complete absence of oxygen is, however, restricted to the extension of the coleoptile and a limited growth of the seminal roots. If anoxia continues beyond this stage, further development is arrested, the coleoptile becomes highly epinastic and the plant dies (Vartapetian, Andreeva & Nuritdinov, 1978). Under conditions of strict anoxia only one other species has been reported as showing some development, namely barnyard grass (*Echinochloa crus-galli*). This species is a common weed of rice paddy fields and shows coleoptile extension under anoxia but not root emergence as with rice (Kennedy *et al.*, 1980). The rhizomes shown in Figs. 1 and 2 have an anoxia endurance-capacity far greater than that of rice or barnyard grass in that they can be kept alive under strictly anaerobic conditions for up to two months. *Scirpus maritimus* has remained healthy and been able to produce under anoxia, normal shoot growth from unextended lateral buds after eight weeks of anaerobic incubation at 22 °C (Barclay & Crawford, 1982).

Table 1. *Shoot extension and survival of wetland species kept under total anoxia in an anaerobic incubator at 22 °C for 7 days*

(*a*) Species surviving and showing sustained shoot extension under anoxia
Schoenoplectus lacustris (L.) Palla
Schoenoplectus tabernaemontani (C.C. Gmel.) Palla
Scirpus maritimus L.
Typha angustifolia L.
Spartina anglica C. E. Hubbard
Potamogeton filiformis Pers.

(*b*) Species surviving but showing no shoot extension under anoxia
Phragmites australis (Cav.) Trin. ex Steudel
Phalaris arundinacea L.
Iris pseudacorus L.
Eleocharis palustris (L.) Roem. & Schult.
Filipendula ulmaria (L.) Maxim.
Cyperus papyrus L.
Cyperus alternifolius L.

(*c*) Species killed by anoxia
Glyceria maxima (Hartm.) Holmberg.
Juncus effusus L.
Juncus conglomeratus L.
Ranunculus lingua L.
Mentha aquatica L.
Oryza sativa L. var. Oeiras
Cyperus diffusus Wahl

The incubating gas (85% nitrogen, 10% hydrogen and 5% carbon dioxide) was circulated over a palladium catalyst to remove traces of oxygen and at all times the degree of anaerobiosis was sufficient to keep a solution of methylene blue in the reduced colourless state.

Fig. 1. Rhizomes of *Schoenoplectus lacustris* showing growth of the terminal bud and distal shoot which had taken place during incubation under total anoxia at 22 °C for 15 days.

Fig. 2. Rhizomes of *Scirpus maritimus* showing growth of terminal bυ d which had taken place during incubation under total anoxia at 22 °C for 7 days.

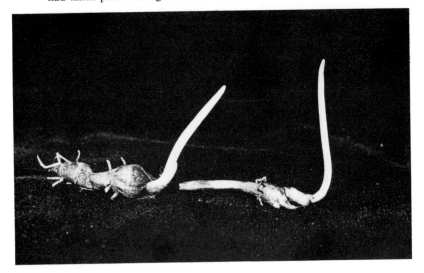

Metabolic activity under anoxia

Under natural conditions the removal of oxygen is usually accompanied by an increase in carbon dioxide. It has often been suggested, but with little clear experimental evidence, that carbon dioxide may be a regulator of plant metabolism under anaerobic conditions (Rowe & Beardsell, 1973; Zemlianukhin & Ivanov, 1978). Fig. 3 shows the effect of accumulating carbon dioxide on anaerobic carbon dioxide evolution in a number of species. The greatest reduction in carbon dioxide evolution is found in *Schoenoplectus lacustris* and *Iris pseudacorus*. This reduction in anaerobic respiration is not due to ageing because the rate of carbon dioxide evolution can be restored

Fig. 3. Effect of increasing carbon dioxide on the anaerobic output of carbon dioxide in (*a*) rhizomes of *Schoenoplectus lacustris*, (*b*) rhizomes of *Iris pseudacorus*, (*c*) rhizomes and roots of *Phragmites australis*, (*d*) rhizomes and roots of *Glyceria maxima*. The respiring organs were kept in a cuvette on moist filter paper under nitrogen and the natural increase in carbon dioxide content was monitored by infrared gas analysis in a closed system at 20 °C.

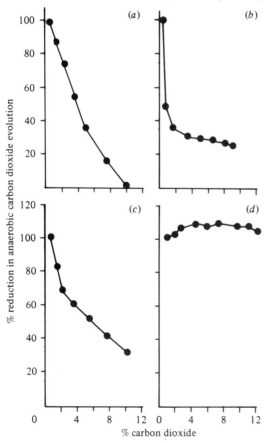

if the carbon dioxide is removed from the gas phase. It is possible that the production of some other volatile compound such as ethylene could be affecting the rate of carbon dioxide production. This, however, is unlikely, as gassing the rhizomes with nitrogen containing 5% carbon dioxide from a cylinder is equally inhibitory to carbon dioxide evolution.

The reduction in carbon dioxide production under high concentrations of carbon dioxide also reduces the production of ethanol. Incubation of *Iris pseudacorus* rhizomes in nitrogen containing 0, 5 and 10% carbon dioxide for one week causes the ethanol level to drop from 85 to 50 and 30 μmol per gram fresh weight respectively. The aerobic control was found to contain 35 μmol ethanol per gram fresh weight. Previous studies have shown that aerobic respiration in *I. pseudacorus* (Brown, Boulter & Coult, 1968) is inhibited by carbon dioxide as in other tissues through its action on succinic dehydrogenase (Bendall, Ranson & Walker, 1960). These present observations extend this inhibitory activity of carbon dioxide to anaerobic metabolism. Although the number and type of organs tested so far are too few to make any predictions as to the ecology of the species which show this type of behaviour, they show the importance of simulating natural conditions with realistic concentrations of carbon dioxide when investigating the effects of flooding on anaerobic metabolism.

A restriction in metabolic activity as seen in lower anaerobic respiration rates and a reduction in the accumulation of ethanol has already been observed in various flood-tolerant roots and seeds. Non-tolerant species usually show the opposite effect, with accelerated glycolytic rates and an increased production of ethanol after flooding (Crawford, 1978).

Metabolic products of anaerobiosis

In microorganisms and lower animals able to withstand lengthy periods of anoxia, mixed fermentations produce a variety of end-products (Hochachka, 1980). The best-documented alternative to ethanol as an end-product of glycolysis in higher plants is lactic acid (Brown, Dufred & Hill, 1969; Leblova, 1978). Malic acid accumulates in concentrations greater than lactate during the hypoxic stage of germination in chickpea seeds (Aldasoro & Nicolas, 1980), as well as in the roots of some flood-tolerant plants when they are inundated (Crawford & Tyler, 1969; Linhart & Baker, 1973). This accumulation of malate is not found when detached roots are incubated under nitrogen (Smith & ap Rees, 1979) and its role as an alternative to ethanol in glycolysis is therefore not proven. Other substances which could permit the anaerobic cycling of NADH to NAD have also been found to accumulate in wetland species. These include shikimic acid, glycerol, glycolic and succinic acid as well as a number of amino acids (Crawford, 1978). It has not been shown as yet whether these substances accumulate at rates which match the anaerobic activity of the plants. This problem is complicated by the observation made above that under natural conditions carbon dioxide accumulation would be expected to cause anaerobic respiration rates in some species to be extremely low.

Conclusions

Among flood-tolerant plants there are species that will survive anoxia and others that will not. In the species able to survive anoxia, anaerobic respiration not only sustains life but in some cases supports the development of new shoots. A characteristic feature of flood-tolerant species able to withstand anoxia is the control of anaerobic respiration with a limitation to the production of ethanol. Under natural conditions carbon dioxide can be expected to have an important inhibitory function in some species in limiting glycolysis. Observations on a range of seeds, tubers and roots show that an acceleration of glycolytic activity under anoxia is usually associated with flood-intolerant species or those that only have to withstand short periods of anaerobiosis.

This research was supported by a grant from the Natural Environment Research Council, which is gratefully acknowledged.

References

Aldasoro, J. & Nicolas, G. (1980). Fermentative products and dark CO_2 fixation during germination of seeds of *Cicer arietinum*. *Phytochemistry*, **19**, 3–5.

Armstrong, W. (1980). Aeration in higher plants. *Advances in Botanical Research*, **7**, 226–33.

Barclay, A. M. & Crawford, R. M. M. (1982). Plant growth and survival under strict anaerobiosis. *Journal of Experimental Botany*, **33**, 541–9.

Bendall, D. S., Ranson, S. L. & Walker, D. A. (1960). Effects of carbon dioxide on the oxidation of succinate and reduced diphosphopyridine nucleotide by *Ricinus* mitochondria. *Biochemical Journal*, **76**, 221–5.

Blackman, F. F. (1928). Analytic studies in plant respiration. III. Formulation of a catalytic system for the respiration of apples and its relation to oxygen. *Proceedings of the Royal Society of London, Series B*, **103**, 491–523.

Brändle, R. (1980). Die Überflutungstoleranz der Seebinse (*Schoenoplectus lacustris* L. Palla). II. Übersicht über die verschiedenen Anpassungs-strategien. *Vierteljahresschrift der Naturforschenden Gesellschaft in Zurich*, **125**, 177–85.

Brown, A., Boulter, D. & Coult, D. A. (1968). The influence of carbon dioxide on the metabolism of rhizome tissue in *Iris pseudacorus*. *Physiologia Plantarum*, **21**, 271–81.

Brown, J. M., Dufred, H. A. & Hill, C. F. (1969). Respiratory metabolism in mangrove seedlings. *Plant Physiology*, **44**, 287–94.

Chirkova, T. V. (1978). Some regulatory mechanisms of plant adaptation to temporal anaerobiosis. In *Plant Life in Anaerobic Environments*, ed. D. D. Hook & R. M. M. Crawford, pp. 139–54. Michigan: Ann Arbor.

Crawford, R. M. M. (1978). Metabolic adaptations to anoxia. In *Plant Life in Anaerobic Environments*, ed. D. D. Hook & R. M. M. Crawford, pp. 119–36. Michigan: Ann Arbor.

Crawford, R. M. M. & Baines, M. (1977). Tolerance of anoxia and ethanol metabolism in tree roots. *New Phytologist*, **79**, 519–26.

Crawford, R. M. M. & Tyler, P. D. (1969). Organic acid metabolism in relation to flooding tolerance in roots. *Journal of Ecology*, **57**, 235–44.

Hochachka, P. W. (1980). *Living without Oxygen*. Cambridge, Mass.: Harvard University Press.

Hook, D. D. & Brown, C. L. (1973). Root adaptation and relative flood tolerance of five hardwood species. *Forest Science*, **19**, 225–9.

Hook, D. D. & Crawford, R. M. M. (eds.) (1978). *Plant Life in Anaerobic Environments*. Michigan: Ann Arbor.

Kennedy, R. A., Barrett, S. C. H., Vander Zee, D. & Rumpho, M. E. (1980). Germination and seedling growth under anaerobic conditions in *Echinochloa crus-galli* (barnyard grass). *Plant Cell and Environment*, **3**, 243–8.

Leblova, S. (1978). Pyruvate conversions in higher plants during natural anaerobiosis. In *Plant Life in Anaerobic Environments*, ed. D. D. Hook & R. M. M. Crawford, pp. 155–68. Michigan: Ann Arbor.

Linhart, Y. B. & Baker, G. (1973). Intra-population differentiation of physiological responses to flooding in a population of *Veronica peregrina* L. *Nature, London*, **242**, 275.

Öpik, H. (1980). *The Respiration of Higher Plants*. Studies in Biology 120. London: Edward Arnold.

Pradet, A. & Bomsel, J. L. (1978). Energy metabolism in plants under hypoxia and anoxia. In *Plant Life in Anaerobic Environments*, ed. D. D. Hook & R. M. M. Crawford, pp. 89–118. Michigan: Ann Arbor.

Raymond, P., Bruzau, F. & Pradet, A. (1978). Etude du transport d'oxygène des parties aériennes aux racines à l'aide d'un Paramètre de métabolisme: la charge enérgétique. *Comptes rendus de l'Académie des Sciences, Paris*, **286**, 1061–3.

Rowe, R. N. & Beardsell, D. V. (1973). Waterlogging of fruit trees. *Horticultural Abstracts*, **43**, 534–48.

Sherwin, T. & Simon, E. W. (1969). The appearance of lactic acid in *Phaseolus* seeds germinating under wet conditions. *Journal of Experimental Botany*, **20**, 776–85.

Smith, A. M. & ap Rees, T. (1979). Pathways of carbohydrate fermentation in the roots of marsh plants. *Planta*, **146**, 327–34.

Turner, J. S. (1951). Respiration (the Pasteur effect in plants). *Annual Review of Plant Physiology*, **2**, 145–68.

Vartapetian, B. B., Andreeva, I. N. & Nuritdinov, N. (1978). Plant cells under oxygen stress. In *Plant Life in Anaerobic Environments*, ed. D. D. Hook & R. M. M. Crawford, pp. 13–88. Michigan: Ann Arbor.

Zemlianukhin, A. A. & Ivanov, B. F. (1978). Metabolism of organic acids of plants in conditions of hypoxia. In *Plant Life in Anaerobic Environments*, ed. D. D. Hook & R. M. M. Crawford, pp. 203–68. Michigan: Ann Arbor.

ANDRÉ BERVILLÉ & MICHÈLE PAILLARD

7   The effect of *Helminthosporium maydis* race T toxin on plant respiration

Cytoplasmically inherited male sterile mutants of grain crops play an important role in modern plant breeding. The Texas male-sterile line (TMS) is very susceptible to Southern corn leaf blight (*Helminthosporium maydis*, race T (HMT)).

Bervillé & Demarly (1970) proposed that male sterility was associated with mutations in the mitochondrial system. The mutation appears to lead to a deficiency in energy conversion such that the microspores cannot ripen, thus resulting in male sterility. It is therefore possible that the sensitivity of TMS plants to HMT toxin resides in the alteration of the mitochondrial structure. We have been studying the structure and mode of action of HMT toxin in the hope that such knowledge will allow the improvement of TMS cytoplasm for resistance to HMT and reveal new knowledge concerning the physiological basis for male sterility. If cytoplasmic male sterility and susceptibility to HMT prove to be linked, then plant breeders will have to find new ways of inducing male-sterile mutants which are resistant to HMT.

Experiments using HMT toxin, *Phyllostycta maydis* toxin and methomyl (Lannate) show that they all have a similar specificity for TMS cytoplasm and have similar effects on TMS mitochondria (see Scheffer, 1977, for references).

Experiments with HMT toxin show inhibition of carbon dioxide fixation through effects on stomatal aperture, whereas the respiration of TMS leaves and coleoptiles, but not roots, increased after toxin treatment. HMT toxin inhibited dark carbon dioxide fixation but had no effect on the activity of phosphoenolpyruvate carboxylase. In whole cells the HMT toxin stimulated the leakage of electrolytes including phosphate and caused the depolarisation of the transmembrane electrical potential. Work with protoplasts has revealed that HMT toxin prevents increases in volume and leads to eventual collapse. Ultrastructural studies show that HMT toxin causes specific mitochondrial damage inside the protoplast and analytical studies reveal that this is accompanied by a decrease in the ATP concentration. It should be emphasised that there are no effects of HMT toxin on normal plants or cells.

The influence of HMT toxin on isolated mitochondria was first reported by Miller & Koeppe (1971). The toxin has been found to inhibit the oxidation of NAD$^+$-linked substrates in isolated mitochondria. The influence of the toxin on succinate oxidation varies from slight inhibition to slight stimulation depending on the composition of the assay medium, the corn lines or toxin isolate. The oxidation of external NADH is strongly stimulated by the toxin; this we have labelled as a stimulation coefficient (SC in Fig. 1). Another major effect of HMT toxin on isolated mitochondria is to cause an increase in light transmittance, due to swelling, both in the absence and presence of an electron donor.

The HMT toxin does not have any effect on normal corn mitochondria or on corn male-sterile cytoplasms other than TMS. No effects of HMT toxin have been observed on mitochondria isolated from a variety of other plant and animal tissues (A. Bervillé & M. Paillard, unpublished). However, there are reports that HMT toxin stimulates the activity of several mitochondrial enzymes, i.e. ATPase, cytochrome oxidase and succinate–cytochrome c reductase.

This survey raises several questions:

    1. Is there a specific site of action in the TMS mitochondrial respiratory chain which could account for all the observations?

    2. If the mitochondria are not involved in the mechanism of cytoplasmic male sterility, how is it then possible to explain male sterility?

Fig. 1. Effect of methomyl on T mitochondrial respiration rates and coupling with NADH as respiratory substrate. Inhibitory effect on the rate of oxygen uptake of rotenone added after methomyl. Assay conditions: mannitol, 0.3 M; KCl, 10 mM; MgCl$_2$, 5 mM; KH$_2$PO$_4$, 10 mM; pH 7.2 with KOH; bovine serum albumin, 0.1%. Additions along the traces according to the arrows: M, about 1.4 mg protein of a mitochondrial preparation as described by Bervillé & Cassini (1974); NADH, 1.5 mM; ADP, 150 μmol; methomyl as Lannate, 4.1 mM. Rotenone is dissolved in dimethyl sulphoxide (DMSO); 1 μmol is added in (*b*) and (*d*) for 1.6 × 10^{-3} M and 5 μl of pure DMSO is added in the control (*a*). The 2,4-dinitrophenol (DNP) is dissolved in methanol; 10 μl is added for a final concentration of 10^{-5} M. Oxygen uptake is expressed as nmol O$_2$ mg^{-1} protein min^{-1} in a 3 ml cell. IE, inhibition efficiency for rotenone (%); SC, stimulation coefficient for methomyl.

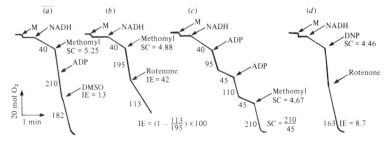

3. What is the primary site of HMT toxin effect in the cell: the plasma membrane, the chloroplast, the mitochondrion, or somewhere else?
4. Could the mechanism of HMT toxin action explain the effects on both the plasma membrane and the mitochondrion? An increase in permeability of the mitochondrial membranes has been proposed (Gregory, Earle & Gracen, 1977; Ber,illé, 1979).

There is experimental evidence to support the hypothesis that HMT toxin increases the permeability of the mitochondrial membrane (Matthews, Gregory & Gracen, 1979), and new evidence is presented in Figs. 1 and 2. From the data in Fig. 1 it can be seen that the addition of methomyl (which has a similar effect to HMT toxin) causes a strong uncoupling of NADH oxidation and induces some sensitivity to rotenone which suggests that some of the external NADH is being oxidised by the internal rotenone-sensitive NADH dehydrogenase. The data in Fig. 2 show that methomyl strongly inhibits the oxidation of malate (trace b). The inhibition can be reversed by adding NAD^+; this suggests the inhibition is not at the level of the respiratory chain and is consistent with the view that the toxin causes the loss of NAD^+ from the mitochondrial matrix. The ability of methomyl to uncouple oxidative phosphorylation can be seen in traces (b) and (c) of Fig. 1. Moreover, unpublished results of A. Bervillé & Ghazi show that methomyl collapses the proton-motive force by either increasing the proton permeability or inhibiting the ejection of protons. The latter suggestion appears more plausible after checking the permeability of the mitochondrial membrane using acid pulse techniques.

These results suggest that the HMT toxin may be toxic to TMS plants by interacting with the cell membrane, which it must cross before gaining access to the mitochondria and interfering with an energy-dependent electrogenic mechanism resulting in the depolarisation of transmembrane electric potential

Fig. 2. Effect of methomyl on T mitochondrial respiration rates and coupling with malate as respiratory substrate. Restoration of malate oxidation by NAD^+ in the presence of methomyl. Assay reaction medium as in Fig. 1, except: malate, 30 mM; ADP, 225 μM; NAD, 0.2 mM. About 3 mg of protein are added for each trace.

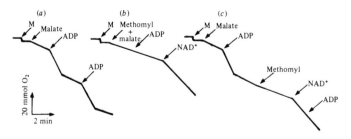

differences. Ion-specific ATPases may be candidates for the site of action of the toxin. However, no direct effect of HMT toxin on any cell membrane has been reported and no component of the cell membrane is known to be coded by cytoplasmic genes. In addition, treatment of TMS protoplasts with the toxin caused the mitochondria to be disrupted after 30 minutes (Gregory, Earle & Gracen, 1980), while no damage could be seen to the cell membrane until the treatment had lasted between one and two days.

References

Bervillé, A. (1979). Etude préliminaire de la stérilité mâle cytoplasmique. Doctoral thesis, Université de Paris Sud, Orsay.

Bervillé, A. & Cassini, R. (1974). Effets comparatifs de la toxine produite par *Helminthosporium maydis* race T sur des mitochondries isolées de maïs à cytoplasme, mâle-stérile 'Texas' et à cytoplasme normal. *Comptes Rendus de l'Académie des Sciences, Paris, Series D*, **278**, 885–8.

Bervillé, A. & Demarly, Y. (1970). La stérilité mâle cytoplasmique chez les végétaux. In *Comptes Rendus du Congrès Eucarpia, Section Plantes Horticoles, Versailles*.

Gregory, P., Earle, E. & Gracen, V. E. (1977). Biochemical and ultrastructural aspects of southern corn leaf blight disease. In *Host Plant Resistance to Pests*, ed. P. A. Hedin, ACS Symposium Series 62, pp. 90–114. Washington, DC: American Chemical Society.

Gregory, P., Earle, E. & Gracen, V. E. (1980). Effects of purified *Helminthosporium maydis* race T toxin on the structure and function of corn mitochondria and protoplasts. *Plant Physiology*, **66**, 477–81.

Matthews, D. E., Gregory, P. & Gracen, V. E. (1979). *Helminthosporium maydis* race T toxin induces leakage of NAD^+ from T cytoplasm corn mitochondria. *Plant Physiology*, **63**, 1149–53.

Miller, R. J. & Koeppe, D. E. (1971). Southern Corn leaf blight: susceptible and resistant mitochondria. *Science*, **173**, 67.

Scheffer, R. (1977). Host specific toxins in relation to pathogenesis and disease resistance. In *Encyclopedia of Plant Physiology, New Series*, ed. R. Heitefuss & P. M. Williams, pp. 247–69. Berlin: Springer-Verlag.

8 The influence of herbicides on the fluidity of the mitochondrial membrane

The action of herbicides on the fluidity of plant membranes was neglected until 1978 (Moreland, 1980). (Lenaz, Curatola & Masotti (1975) have reviewed many aspects of the perturbation of membrane fluidity.) Plant mitochondrial membranes are the only system so far used to study the effects of herbicides on the fluidity of membranes in plant cells. This is not because all herbicides are necessarily harmful to mitochondria but because mitochondria are easy to isolate and characterise. For this reason mitochondrial membranes constitute a model system and effects of herbicides observed here may indicate similar effects in other plant cell membranes.

We have divided this paper into four parts: (1) the definition of membrane fluidity and the biological consequences of its variation, (2) the techniques that may be used to study herbicide actions on membrane fluidity, (3) results, (4) the significance of these results for the mode of action of herbicides.

Membrane fluidity

Biological membranes are thought to exist primarily as lipid bilayers containing proteins embedded to various depths. The presence of the proteins in the bilayer affects the mobility of neighbouring lipids (Rumsby, 1979). The plant mitochondrial inner membrane contains 80% protein (Dizengremel & Kader, 1980) and, even if the proteins are relatively aggregated, the phospholipid bilayer probably forms only a small part of the total membrane area. Thus, while mitochondria may be the easiest membrane system to study they may not be the most representative. Normally, membrane lipids retain the rotational and diffusional mobility described qualitatively by 'fluidity'. Viscosity would be the ultimate physical parameter to characterise mobility but viscosity measurements which give the 'average viscosity' may be unreliable in such a situation (Hare, Amiell & Lussan, 1979) and do not indicate fluidity variations across or within the membrane. Thus, the use of 'viscosity' is no more precise than 'fluidity'.

Membrane fluidity affects membrane functions as follows:

1. Transport activities are reduced when fluidity is decreased.

2. Variations in temperature change the activation energy (E_a) of membrane-bound enzymes by acting either directly on the enzymic protein or via the physical state of the lipid phase (Quinn & Williams, 1978). Variation in E_a might affect metabolic pathways involving both soluble and membranous components, such as oxidative metabolism (Raison, 1973), to give an imbalance in their rate and regulation.

3. The extent of contact between the surrounding medium and a membrane protein is a function of the hydrophilic part of the protein and the fluidity of the membrane. In erythrocyte membranes, increased fluidity increases exposure of the proteins to the medium (Borochov & Shinitzky, 1976).

4. The conformation of membrane proteins can be influenced by changes in the composition and fluidity of the membrane lipids. The helical content of the Folch–Lees myelin protein is greater in a lipid environment than in water (see Lenaz *et al.*, 1975).

5. Various kinetic parameters of enzymic membrane proteins can also be affected by the lipid fluidity; numerous examples are given by Lenaz *et al.* (1975).

Several exogenous agents are known to affect membrane fluidity, including calcium ions (decrease fluidity), detergents (increase fluidity), anaesthetics (increase fluidity) and alcohols (variable actions) (Lenaz *et al.*, 1975).

Techniques

Two techniques have been used to assess changes in membrane fluidity after addition of a herbicide to plant mitochondria: swelling and fluorescence depolarisation.

Swelling

Mitochondria behave as osmometers, changing their volume according to the relative osmotic potentials of matrix and suspending medium. Plant mitochondria suspended in isotonic potassium chloride do not swell because potassium ions cannot penetrate the inner membrane, but if valinomycin (a potassium ionophore) is added, potassium can enter; as the inner membrane is relatively permeable to chloride (Moore & Wilson, 1977), chloride ions follow potassium into the matrix. This salt entry causes the entry of water and swelling occurs at a rate limited either by chloride or potassium/valinomycin permeation, depending on the valinomycin concentration. In the latter case swelling depends on valinomycin movements and thence on fluidity. Conversely, gramicidin makes the membrane permeable to potassium ions (and protons) by forming pores in the membrane, a process which is independent of membrane fluidity. Thus the action of a chemical on the valinomycin- and/or gramicidin-induced swelling in iso-osmolar potassium chloride can be used to assess its effect on mitochondrial membrane fluidity. Swelling in ammonium phosphate can also be used to measure fluidity. Ammonia enters the matrix to form ammonium and hydroxyl ions. The hydroxyl is then exchanged for the external phosphate by the fluidity-dependent phosphate/hydroxyl antiporter. Like swelling in gramicidin, swelling in ammonium acetate involves only diffusion and is independent of fluidity.

Swelling is conveniently monitored by variations in light scattering or absorption at 520 nm.

Fluorescence depolarisation

In the technique of fluorescence depolarisation a fluorescent probe incorporated into a membrane is illuminated with polarised light (at the absorbing wavelength of the probe) and the polarisation of the fluorescence emitted by the probe analysed. Rotation of the probe causes depolarisation to an extent depending on the rotational freedom of the probe. Fluorescence depolarisation (P) is usually expressed as:

$$P = \frac{T_{\parallel} - T_{\perp}}{T_{\parallel} + T_{\perp}}$$

where $T_{\parallel}$ and $T_{\perp}$ are fluorescence intensities parallel and perpendicular to initial polarisation. Light scattering by the mitochondria can also cause depolarisation and must be minimised by using short light paths in the cuvette and analysing re-emitted light from a directly illuminated area of the cuvette. Mitochondrial suspensions also fluoresce, and sufficient probe must be incorporated to overcome this effect.

Other techniques

Techniques which would yield useful information but which have not yet been used in studies of herbicides are: differential scanning calorimetry, X-ray diffraction, nuclear magnetic resonance, and electron spin resonance using spin-labelled probes (Lenaz *et al.*, 1975).

Results

Little is known about the effect of herbicides on mitochondrial membrane fluidity. Moreland & Huber (1978, 1979) made a survey of several herbicide families, using the swelling technique to monitor fluidity changes. They found that substituted uracils, *s*-triazines, pyridazinones and phenylureas had no effect on membrane fluidity in mung bean mitochondria. On the other hand, carbamates (Fig. 1), phenylamides and dinitroanilines (Fig. 1) decreased fluidity when used at concentrations of 10^{-5} to 10^{-4} M. This study was initiated to explain why the herbicides inhibited several sites in the mitochondrial electron transport chain. Such multiple inhibition illustrates the multi-site effects that can result from membrane fluidity changes.

We have used the swelling technique with carbamate herbicides and potato tuber mitochondria (C. Gauvrit, unpublished). Propham, Chlorpropham, Swep and Phenmedipham all decreased fluidity.

Moreland & Huber (1978) found that Chlorpropham and Dinitramine increased fluorescence depolarisation of the probe 1,6-diphenylhexatriene in mung bean mitochondria. In our laboratory we failed to reproduce these results on liposomes (artificial membranes) and potato tuber mitochondria (J. M. Ducruet & C. Gauvrit, unpublished). An explanation could be that the two techniques do not give the same kind of information about the membrane – or the physical state and sensitivity to herbicides of membranes could vary according to their source. The valinomycin test indicates the hindrance encountered by valinomycin in moving through the membrane, but the ionophore crosses regions of varying fluidity, the hydrocarbon core being more fluid than the surface. The fluorescent probe test indicates variations in fluidity in the hydrocarbon core where the probe is located, and this may not be in the same location as the herbicide. Additionally the physical state

Fig. 1. Chemical formulas of the herbicides discussed in the text: (*a*) carbamates, (*b*) dinitroanilines.

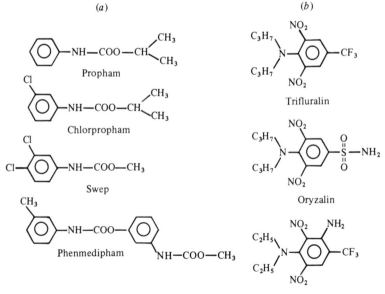

of mung bean membranes differs greatly from those of potato and liposomes, the mung bean membranes being far more fluid ($P = 0.040$: Moreland & Huber, 1978) than the latter ($P = 0.25$: J. M. Ducruet & C. Gauvrit, unpublished), and it is reasonable to expect differences in their sensitivities to the herbicides.

Membrane fluidity and the mode of action of herbicides

Many herbicides seem to have a single site of action, while others, including carbamates and dinitroanilines, display a variety of biochemical actions. The observation that this latter group decreases mitochondrial membrane fluidity might be extrapolated to many other membranes and indicate interference with various aspects of metabolism. Apparently unrelated actions could all result from a primary action on membrane fluidity. For instance three main actions have been suggested for carbamates and dinitroanilines: inhibition of photosynthesis, inhibition of oxidative phosphorylation and inhibition of tubulin polymerisation (Ashton & Crafts, 1973). It is easy to understand how the first two effects could result from a decrease in fluidity slowing electron transport. Moreover, recent data indicate that the action on tubulin might also depend on membrane fluidity. Hertel *et al.* (1980) have shown that Trifluralin, Oryzalin (dinitroanilines), Chlorpropham and Propham (carbamates) inhibit ATP-driven calcium uptake by maize mitochondria at concentrations ranging from 10^{-6} to 10^{-4} M. This effect might be the result of a decrease in membrane fluidity if a mobile carrier is implicated in calcium uptake. (In these experiments electron transfer inhibition is not involved since ATP is the energy source, although the ATPase may have been inhibited: Moreland & Huber, 1979.) If mitochondria cease to accumulate calcium, the free cytoplasmic concentration could rise and tubulin polymerisation cease.

Conclusion

As new light is shed on the mode of action of herbicides by membrane fluidity studies, the multi-site effects of herbicides such as carbamates and dinitroanilines may be better understood. Even for families of herbicides whose primary site of action is more apparent (such as triazines and substituted ureas) studies of fluidity could explain many side or secondary effects. Mitochondria are a convenient, albeit not particularly representative, tool to use for pioneering studies, but fluidity studies on other membranes such as plasmalemma, tonoplast, endoplasmic reticulum and thylakoid membrane will be required for a fuller understanding of the mechanism of herbicide action.

We are indebted to Dr J. M. Ducruet for helpful discussions during the preparation of the manuscript.

References

Ashton, F. M. & Crafts, A. S. (1973). *Mode of Action of Herbicides*. New York: Wiley.

Borochov, H. & Shinitzky, M. (1976). Vertical displacement of membrane proteins mediated by changes in microviscosity. *Proceedings of the National Academy of Sciences, USA*, **73**, 4526–30.

Dizengremel, P. & Kader, J. C. (1980). Effect of ageing on the composition of mitochondrial membranes from potato slices. *Phytochemistry*, **19**, 211–14.

Hare, F., Amiell, J. & Lussan, C. (1979). Is an average viscosity tenable in lipid bilayers and membranes? *Biochimica et Biophysica Acta*, **555**, 388–408.

Hertel, C., Quader, H., Robinson, D. G. & Marme, D. (1980). Antimicrotubular herbicides and fungicides affect Ca^{2+} transport in plant mitochondria. *Planta*, **149**, 336–40.

Lenaz, G., Curatola, G. & Masotti, L. (1975). Perturbation of membrane fluidity. *Journal of Bioenergetics*, **7**, 223–99.

Moore, A. L. & Wilson, S. B. (1977). Translocation of some anions, cations and acids in turnip (*Brassica napus* L.) mitochondria. *Journal of Experimental Botany*, **28**, 607–18.

Moreland, D. E. (1980). Mechanism of action of herbicides. *Annual Review of Plant Physiology*, **31**, 597–638.

Moreland, D. E. & Huber, S. C. (1978). Fluidity and permeability changes induced in the inner mitochondrial membrane by herbicides. In *Plant Mitochondria*, ed. G. Ducet & C. Lance, pp. 191–8. Amsterdam: Elsevier North-Holland.

Moreland, D. E. & Huber, S. C. (1979). Inhibition of photosynthesis and respiration by substituted 2,6-dinitroaniline herbicides. *Pesticide Biochemistry and Physiology*, **11**, 247–57.

Quinn, P. J. & Williams, W. P. (1978). Plant lipids and their role in membrane function. *Progress in Biophysics and Molecular Biology*, **34**, 109–73.

Raison, J. K. (1973). Temperature-induced phase changes in membrane lipids and their influence on metabolic regulation. *Symposia of the Society for Experimental Biology*, **27**, 485–512.

Rumsby, M. G. (1979). Organisation in biological membranes. In *Companion to Biochemistry*, vol. 2, ed. A. T. Bull, J. R. Lagnado, J. D. Thomas & K. F. Tipton, pp. 161–204. Harlow: Longman.

PART II

The biochemistry of plant respiration

T. ap REES

9 The reactions, location and functions of glycolysis and the oxidative pentose phosphate pathway in higher plants

In this article I tried to meet the editor's request for a brief general account of glycolysis and the pentose phosphate pathway in plants that is aimed at undergraduates rather than the cognoscenti.* I have done this by confining myself to the aspects mentioned in the title.

Reactions

The basic reactions of glycolysis (Fig. 1) need no further comment. There is overwhelming evidence that they operate in plants (Beevers, 1961; Davies, Giovanelli & ap Rees, 1964). This evidence includes the response of respiration to selective inhibitors, the way in which specifically labelled substrates are metabolised by intact tissues, and the demonstration that plants contain the appropriate enzymes in sufficient quantities to catalyse the observed rates of respiration. Further support is provided by the behaviour of the intermediates when the rate of glycolysis is varied; see, for example, Dixon & ap Rees (1980). It is difficult to conceive of a living plant cell which does not depend upon glycolysis.

The reactions of the oxidative pentose phosphate pathway are best considered as two groups, the first of which catalyses the oxidative decarboxylation of glucose-6-phosphate to ribulose-5-phosphate (Fig. 2). The two dehydrogenases involved are specific for NADP, and C-1 of glucose is released as carbon dioxide. These reactions occur in most, if not all, plant cells. A very wide range of tissues have been shown to contain the two dehydrogenases, and to release C-1 of [^{14}C]glucose as $^{14}CO_2$ to a greater extent than C-6. Even where yields of $^{14}CO_2$ from C-1 and C-6 are comparable, the pathway may still be operating but be masked by pentan synthesis. The latter releases C-6 preferentially in amounts which can have a marked effect on the patterns of $^{14}CO_2$ production from [^{14}C]glucose (Stitt & ap Rees, 1978). The second group of reactions of the pentose phosphate pathway, the

* This account was delivered to the volume editor in January 1981.

89

Fig. 1. Glycolysis. The numbers show the distribution of the carbons of glucose and the letters refer to the enzymes: a, hexokinase; b, phosphoglucomutase; c, glucosephosphate isomerase; d, phosphofructokinase; e, fructose-1,6-bisphosphate aldolase; f, triosephosphate isomerase; g, NAD-specific glyceraldehydephosphate dehydrogenase; h, phosphoglycerate kinase; i, phosphoglyceromutase; j, enolase; k, pyruvate kinase; l, pyruvate decarboxylase; m, alcohol dehydrogenase; n, lactate dehydrogenase. FRU, fructose; GLC, glucose; P, phosphate; P_2, bisphosphate.

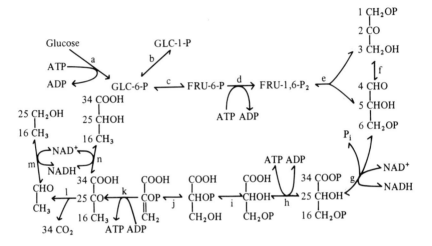

Fig. 2. Conventional representation of the oxidative pentose phosphate pathway. The numbers show the distribution of the carbons of glucose and the letters refer to the enzymes: a, glucose-6-phosphate dehydrogenase; b, 6-phosphogluconate dehydrogenase; c, ribosephosphate isomerase; d, ribulosephosphate-4-epimerase; e, transketolase; f, transaldolase. P, phosphate.

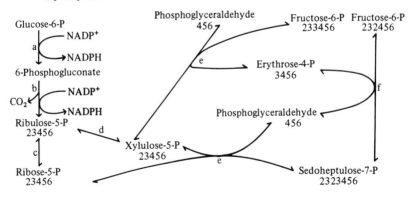

non-oxidative section, converts ribulose-5-phosphate to triose phosphate and hexose-6-phosphate. The presence of the necessary enzymes, and the manner in which specifically labelled substrates are metabolised, leave little doubt that these overall conversions occur in plant cells in general (Beevers, 1961; ap Rees, 1974). However, we do not know the precise pathway taken. This is largely because all of the enzymes involved are readily reversible and many have broad specificities. Thus it is possible to formulate credible schemes which differ in detail from Fig. 2 but give identical labelling of the products. A more pronounced deviation from Fig. 2 has been suggested by Williams (1980), who has proposed that there are two forms of the pathway. One is that in Fig. 2. The other is more complex, involves arabinose-5-phosphate and octulose-1,8-bisphosphate, and requires a pentose-5-phosphate 2'-epimerase and a phosphotransferase. The products of the latter pathway are said to be glucose-6-phosphate (423456), fructose-6-phosphate (653456) and dihydroxyacetone phosphate (232): the numbers represent the positions in the products of the carbons of the original glucose.

The labelling patterns of free and polymerised sugars derived from products of the pentose phosphate pathway have been determined after supplying labelled intermediates to a range of plant tissues (Beevers, 1961). The results of these and of other labelling experiments strongly support the scheme shown in Fig. 2 or some close variant of it (ap Rees, 1974). For Williams's scheme to operate in these tissues, there would have to be very substantial differences between the labelling of the derivatives studied and that of the immediate products of the pathway. There is, as yet, no evidence that such differences exist in plants. It is entirely possible that the alternative pathway operates in some plants but at present there is no convincing evidence that it does. The claim that the alternative pathway does function in plants (Williams, 1980) appears to be based on the argument that it is one of a number of possible explanations of the Gibbs effect in the reductive pentose phosphate pathway (Clark, Williams & Blackmore, 1974) and an unsubstantiated statement that phosphotransferase is present in *Chlorella* (Williams, 1980).

Interrelationship

The relationship between glycolysis and the pentose phosphate pathway is very close as they are not separated from each other and they share enzymes and substrates. Apart from the initial division of glucose-6-phosphate between the two pathways, the central issue is the fate of the fructose-6-phosphate and the phosphoglyceraldehyde formed by the pentose phosphate pathway. The former may enter glycolysis, be converted to oligosaccharides or polysaccharides, or be recycled through the pentose phosphate pathway.

The phosphoglyceraldehyde can enter glycolysis in all instances, but we would not expect it to be recycled, or be converted to oligosaccharides or polysaccharides, unless there is present a fructose-1,6-bisphosphatase to permit the synthesis of fructose-6-phosphate from triose phosphate. Fructose-1,6-bisphosphatase has been regarded as being confined to photosynthetic and gluconeogenic cells of plants, but it now appears that it is more widespread as there is evidence for its presence in maize and pea roots (Harbron, Foyer & Walker, 1981) and in suspension cultures of soybean (F. D. Macdonald & T. ap Rees, unpublished). Nonetheless, the results of labelling experiments, whilst establishing that recycling of phosphoglyceraldehyde occurs in chloroplasts, suggest that in most other instances it enters glycolysis directly.

The fate of the fructose-6-phosphate formed in the pentose phosphate pathway is indicated by the manner in which plant tissues metabolise [1-^{14}C]glucose and [6-^{14}C]glucose. The pathway leads to an excess of label from C-1 over that from C-6 in carbon dioxide, and a corresponding excess of label from C-6 over that from C-1 in ribulose-5-phosphate and compounds derived from it. The fate of the above ribulose-5-phosphate will be revealed by determining which compounds are more heavily labelled by [6-^{14}C]glucose than by [1-^{14}C]glucose. In carrot slices (Table 1) the contribution of [1-^{14}C]glucose to carbon dioxide exceeded that from [6-^{14}C]glucose by 15% of the absorbed label. The sum of the contributions of [6-^{14}C]glucose to organic acids, amino acids, proteins and lipids exceeded that from [1-^{14}C]glucose by 13% of the absorbed label. Thus most of the ribulose-5-phosphate formed by the oxidative reactions of the pathway was converted to the above

Table 1. *Distribution of ^{14}C after supplying specifically labelled glucose to slices of carrot storage tissue*

Fraction	Percentage of absorbed ^{14}C recovered per fraction after supplying:	
	[1-^{14}C]glucose	[6-^{14}C]glucose
Carbon dioxide	21	6
Organic acids	5	10
Lipids	3	5
Amino acids	10	14
Proteins	5	7
Sugars[a]	44	48
Polysaccharides	8	7

[a] Most of the label in sugars was in sucrose.
From ap Rees & Beevers (1960).

compounds, which are largely derived from pyruvate. Similar experiments with other tissues and other substrates provide strong evidence that the bulk of the products of the pentose phosphate pathway is metabolised via glycolysis (Fig. 3).

The above view is supported by comparisons of the patterns of $^{14}CO_2$ production from specifically labelled glucose and pyruvate (Fig. 4). The

Fig. 3. The relationship between glycolysis and the oxidative pentose phosphate pathway in the cytosol. P, phosphate; P_2, bisphosphate.

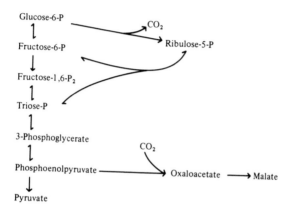

Fig. 4. $^{14}CO_2$ production by slices of carrot storage tissue supplied with specifically labelled substrates. (*a*) [^{14}C]glucose: C-1, ●---●; C-2, ○——○; C-3,4, ●——●; C-6; ○---○. (*b*) [^{14}C]pyruvate: C-1, ●——●; C-2, ○——○; C-3, ○---○. (Data from ap Rees & Beevers, 1960.)

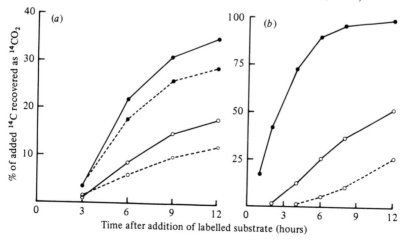

patterns are consistent with the proposal, and two particularly important predictions are met. First, the initial excess of glucose C-1 over C-6 is not followed by a similar excess of C-6 over C-1. This indicates that there is not much recycling in the pentose phosphate pathway. Second, as expected from Figs. 1–3, the yields from glucose C-3,4, C-2 and C-5 are closely comparable to those from pyruvate C-1, C-2 and C-3, respectively.

Some of the products of the pentose phosphate pathway escape glycolysis. Pentose phosphate and erythrose phosphate may be withdrawn from the pathway and used for the synthesis of nucleic acids and phenylpropanoid compounds, respectively. [6-^{14}C]glucose sometimes labels sucrose (Table 1) and polysaccharide more heavily than does [1-^{14}C]glucose. This suggests that some of the fructose-6-phosphate formed in the pentose phosphate pathway is converted via glucose-6-phosphate and glucose-1-phosphate to sugar nucleotides. This suggestion is confirmed by the fact that pentoses fed to plant tissues label sucrose, and the hexose units of polysaccharides (Beevers, 1961).

The conversion of some of the fructose-6-phosphate formed in the pentose phosphate pathway to glucose-6-phosphate suggests that some of the latter re-enters the pathway and thus gives rise to recycling. Evidence that recycling occurs but is restricted is provided in Table 2. The ratio of C-1 to C-6 in $^{14}CO_2$ when [^{14}C]fructose was fed is well below the value of unity expected from glycolysis (Fig. 1) but nothing like as low as that found with [^{14}C]glucose. This strongly suggests that fructose-6-phosphate does enter the pentose phosphate pathway but does so much less readily than does glucose-6-phosphate. The precise division of ribulose-5-phosphate between the above fates will vary with the tissue.

Table 2. $^{14}CO_2$ *production from specifically labelled* [^{14}C]*glucose and* [^{14}C] *fructose supplied to plant tissues*

Tissue	Time from addition of [^{14}C]hexose (min)	% of added ^{14}C-labelled sugar recovered as $^{14}CO_2$			
		[^{14}C]glucose		[^{14}C]fructose	
		C-1	C-6	C-1	C-6
Carrot storage tissue	150	31	9	16	8
Turnip storage tissue	180	32	12	14	8
Pumpkin mesocarp	180	30	8	21	10

From ap Rees *et al.* (1965).

Products

Under aerobic conditions pyruvate is the major, but not the sole, product of glycolysis and the pentose phosphate pathway. An appreciable fraction of the phosphoenolpyruvate formed during carbohydrate oxidation is metabolised by phosphoenolpyruvate carboxylase to oxaloacetate and thence to malate and other acids (Davies, 1979; ap Rees, 1980a; ap Rees, Fuller & Green, 1981). Although the complete significance of this diversion is not yet apparent, it is clear that it plays a number of important roles. Carboxylation of phosphoenolpyruvate to give oxaloacetate is needed to replenish the tricarboxylic acid cycle during the synthesis of amino acids, during the accumulation of organic acids in the maintenance of charge balance, and possibly in the control of pH.

Under anaerobic conditions, where an almost complete dominance of glycolysis over the pentose phosphate pathway would be expected, a variety of products is formed. Although too few species have been examined in enough detail to permit an authoritative generalisation, it is likely that the main products are ethanol and carbon dioxide. In some tissues, such as carrot roots, these are almost the only products. However, it is more commonly found that ethanol accumulation in anoxia is accompanied by that of other compounds, notably lactate, alanine (Smith & ap Rees, 1979a) and γ-aminobutyrate (Streeter & Thompson, 1972). Lactate is formed by the widely distributed lactate dehydrogenase. The pathways to the amino acids are not definitely known but probably involve transaminations which produce keto-acids which can be used for the oxidation of the NADH formed in glycolysis.

It has been suggested that malate, formed via phosphoenolpyruvate carboxylase, is an important product of fermentation in flood-tolerant plants (Crawford, 1978). This may be so, but there is, as yet, no proof that it is. Such proof requires that malate be shown to accumulate in anoxia, and that this accumulation be shown to represent an appreciable fraction of the total products of fermentation. Unless the latter point is met, measurements of malate are uninterpretable. A detailed study of fermentation in the excised apical 1–2 cm of roots of three species of flood-tolerant plants (Smith & ap Rees, 1979a) and one intolerant plant (Smith & ap Rees, 1979b) showed that they all formed ethanol, carbon dioxide, alanine and lactate, but not malate, as significant products. Two of the species studied are amongst those claimed to accumulate malate – a claim based, at least partially, on experiments with excised roots (Crawford, 1967).

Intracellular location

Early methods for fractionation of plant cells resulted in the appearance of the enzymes of both glycolysis and the pentose phosphate pathway in the supernatant fractions of the extracts, and gave rise to the view that both pathways were located exclusively in the cytosol, the soluble phase of the cytoplasm. The development of more refined techniques has shown that, whilst both sequences are found in the cytosol, most of the steps shown in Figs. 1 and 2 also occur in plastids.

The ease with which cellular organisation and compartmentation can be destroyed makes it extremely difficult to prove that an enzyme is located in the cytosol. I know of no instances in which modern techniques have been used to determine the location of every enzyme of both pathways in the same tissue. Nishimura & Beevers (1979) fractionated the endosperm of germinating castor beans so carefully that there was no significant breakage of organelles as fragile as proplastids. Even under these conditions, a substantial, and except for aldolase, major fraction of each glycolytic enzyme was recovered with the cytosol. The same was true for glucose-6-phosphate and 6-phosphogluconate dehydrogenases, transketolase and transaldolase. The other two enzymes of the pentose phosphate pathway were not studied but there is evidence that a substantial proportion of both of them is in the cytosol in pea roots (Emes & Fowler, 1979). These experiments, and comparable ones exemplified by fractionations of pea leaves (Stitt & ap Rees, 1979) and of the developing endosperm of castor bean (Simcox et al., 1977) provide adequate evidence that the cytosol of plant cells contains significant activities of each of the enzymes of glycolysis and the oxidative pentose phosphate pathway.

It is also difficult to decide whether any of an enzyme is located in a fragile organelle in cells in which most of the enzyme is in the cytosol. The following tests are useful. First, crude preparations of the organelle should contain a greater proportion of the activity of the unfractionated homogenate than is expected from cytoplasmic contamination. The mere appearance of activity in a preparation of organelles is not really interpretable unless we know the degree to which the organelles are contaminated by cytoplasm. Second, the enzyme should be shown to co-purify with the organelle in order to demonstrate that it is associated with the organelle in question and not with some other component of the cell. Finally we need evidence that the enzyme is actually within the organelle and not merely absorbed onto or entangled with it. This can be provided by investigating whether the activity in the preparation of organelles is latent, and whether it is protected from proteases.

The substrates and cofactors used to assay many enzymes do not readily cross the membranes of organelles. Thus, if an enzyme is within an isolated organelle, assay of the enzyme *in vitro* under conditions that keep the organelle intact should give a much lower value than is obtained if the organelles are ruptured before the assay. That is, the enzyme should show latency if it is in the organelle and the substances used in the assay cannot enter the organelle. Protection experiments rely on the observation that proteases do not readily enter intact organelles. If an enzyme is within an organelle then it should be protected from proteases so that incubation of the isolated organelle with an appropriate protease would not greatly reduce the activity of the enzyme. The need for the above tests is demonstrated by the behaviour of hexokinase in extracts of pea leaves (Stitt, Bulpin & ap Rees, 1978). An appreciable proportion of the total activity was found in crude preparations of chloroplasts and much of this co-purified with chloroplasts on a sucrose gradient, yet this activity was neither latent nor protected. This does not preclude the presence of some type of hexokinase in the chloroplast, but it does provide compelling evidence that the activity that we measured in the chloroplast preparations was not within the chloroplasts.

There is now quite decisive evidence that appreciable proportions of both glucose-6-phosphate dehydrogenase and 6-phosphogluconate dehydrogenase in leaves are located in the chloroplast. In young leaves of pea about 40% of these enzymes are in the chloroplast (Stitt & ap Rees, 1979). As chloroplasts also contain the enzymes of the reductive pentose phosphate pathway and transaldolase (ap Rees, 1980b), these organelles have the capacity to catalyse the complete oxidative pentose phosphate pathway including recycling of the triose phosphates. There is equally good evidence that chloroplasts contain phosphofructokinase, though the proportion, 25% of the total in pea leaves (Stitt & ap Rees, 1979), is probably lower than that for the above dehydro-genases. The reductive pentose phosphate pathway provides the enzymes needed to convert fructose-1,6-bisphosphate to 3-phosphoglycerate. Thus there is no doubt that chloroplasts can catalyse glycolysis as far as 3-phosphoglycerate.

There *is* doubt as to whether chloroplasts can convert 3-phosphoglycerate to pyruvate via glycolysis. This doubt arises over whether there is any phosphoglyceromutase in the chloroplast. The best evidence that chloroplasts contain this enzyme is the demonstration that preparations of spinach chloroplasts converted $^{14}CO_2$ to fat and that this conversion was diminished by the presence of unlabelled 3-phosphoglycerate, phosphoenolpyruvate and pyruvate, but not by centrifugation through a gradient of sorbitol (Murphy & Leech, 1977, 1978). These experiments are not entirely conclusive because no evidence was presented that the centrifugation did remove cytoplasmic contaminants and that any phosphoglyceromutase in the preparations was actually within the chloroplast.

Experiments with pea shoots led to the conclusion that the chloroplasts contained about 10% of the enolase and pyruvate kinase but no significant amounts of phosphoglyceromutase (Stitt & ap Rees, 1979). Enolase and pyruvate kinase, but not phosphoglyceromutase, were found in crude preparations of chloroplasts in greater proportions than was the cytoplasmic marker, phosphoenolpyruvate carboxylase (Table 3). The sedimentable activity of the first two enzymes, but not that of phosphoglyceromutase, co-purified with intact chloroplasts, was latent and was protected. Thus the enzymic capacities of isolated chloroplasts from peas show that they can oxidise carbohydrate via the pentose phosphate pathway with complete recycling, and via glycolysis but only as far as 3-phosphoglycerate. The behaviour of isolated chloroplasts accords with these conclusions (Stitt & ap Rees, 1980a). The pattern of $^{14}CO_2$ production from specifically labelled [^{14}C]glucose (Fig. 5), which contrasts markedly with that for intact tissues (Fig. 4), is as expected from the above combination of reactions. Further, when isolated chloroplasts were made to break down starch in the dark at rates comparable to those observed *in vivo*, carbon dioxide and a mixture of 3-phosphoglycerate and triose phosphates were exported from the chloroplast.

It is possible that phosphoglyceromutase was lost from the chloroplasts

Table 3. *Activities of enzymes of carbohydrate oxidation in preparations of chloroplasts from pea shoots*

Enzyme	Activity in chloroplast preparation as:	
	% of activity in unfractionated homogenate	nkat per mg chlorophyll
Glucose-6-phosphate dehydrogenase	13.9±1.0	5.5±0.3
6-Phosphogluconate dehydrogenase	13.8±0.3	8.0±0.3
Phosphofructokinase	8.5±0.02	1.8±0.3
Phosphoglyceromutase	0.5±0.08	2.3±0.7
Enolase	3.0±0.20	6.8±1.3
Pyruvate kinase	3.6±0.70	3.7±0.5
Phosphoenolpyruvate carboxylase	0.6±0.01	0.05±0.02
Ribulose bisphosphate carboxylase	30.1±2.72	6.3±3.0
Glyceraldehydephosphate dehydrogenase (NADP)	28.2±1.40	40.3±6.8

Homogenates of pea shoots were filtered through cotton wool and the filtrate, called the unfractionated homogenate, was centrifuged at 6000 g to give a sediment, called the chloroplast preparation, and a supernatant. Values are means ±s.e. of data from three to six different preparations.
From Stitt & ap Rees (1979).

during their preparation. However, it is difficult to see how this single enzyme could have leaked from chloroplasts which had been shown to retain not only each of the other glycolytic enzymes but also those of both the oxidative and reductive pentose phosphate pathways. Thus, at least for pea leaves, the available evidence strongly suggests that the chloroplasts contain the complete oxidative pentose phosphate pathway, and also glycolysis as far as 3-phosphoglycerate. There is not enough information to assess the significance of the apparent absence of phosphoglyceromutase from chloroplasts, but this is clearly of vital importance in respect of the formation of acetyl CoA for the synthesis of fats and isoprenoids in the chloroplast.

There is evidence that many of the enzymes of the two pathways are present in proplastids. Each of the enzymes of the pentose phosphate pathway has been shown to co-purify with proplastids from pea roots (Emes & Fowler, 1979). Although we still need proof that all these enzymes are in the plastids, it seems likely that proplastids that are engaged in nitrate assimilation contain the complete pentose phosphate pathway. It is important to discover whether they can carry out any of the reactions of glycolysis.

Studies with proplastids involved in fat synthesis have provided substantial evidence that they contain 6-phosphogluconate dehydrogenase, transketolase and transaldolase but lack glucose-6-phosphate dehydrogenase (Simcox *et al.*, 1977; Simcox & Dennis, 1978). Similar results were obtained with the

Fig. 5. $^{14}CO_2$ production from specifically labelled [^{14}C]glucose supplied to chloroplasts isolated from pea shoots. C-1, ●--●; C-2, ○——○; C-3,4, ●——●; C-6, ○- - -○. (Data from Stitt & ap Rees, 1980*a*.)

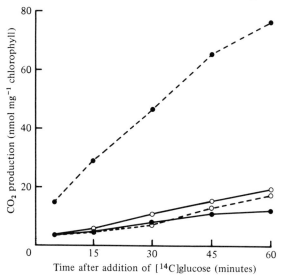

endosperm of germinating castor beans (Nishimura & Beevers, 1979). The absence of glucose-6-phosphate dehydrogenase from the above proplastids is noteworthy as it contrasts with the presence of the enzyme in chloroplasts and implies that 6-phosphogluconate crosses the plastid envelope. There is also compelling evidence that proplastids from developing castor beans contain the complete sequence of glycolytic enzymes from phosphoglucomutase to pyruvate kinase, plus pyruvate dehydrogenase (Dennis & Miernyk, 1982; Miernyk & Dennis, 1982).

Functions

A clear understanding of the functions of glycolysis and the pentose phosphate pathway requires knowledge of their relative activities. I have discussed elsewhere the very considerable problems involved in the measurement of the flux through the two pathways and have argued that we have no reliable precise estimates for higher plants (ap Rees, 1980a). The discovery that many of the reactions of both pathways are almost certainly operating independently in different parts of the same cell complicates these measurements even further. In terms of the cell as a whole, all the available evidence indicates that glycolysis is by far the dominant pathway. It is unlikely that more than 30% of the glucose-6-phosphate metabolised by the cell as a whole enters the pentose phosphate pathway, and it is probable that the usual figure is much lower. As yet, pea leaves are the only tissue for which we have information about the relative activities of the pathways in different parts of the cell. The contribution of the pentose phosphate pathway in the leaf as a whole is almost certainly less than 30% (Stitt & ap Rees, 1978) but in isolated chloroplasts it may be as high as 60% (Stitt & ap Rees, 1980b). Thus the activity of this pathway relative to that of glycolysis is probably much higher in the chloroplasts than in the cytosol. Our present understanding of the control mechanisms whereby the variation in the relative activities of the two pathways is achieved has been discussed recently (ap Rees, 1977, 1980b; Turner & Turner, 1980).

The role of glycolysis is clear. Under aerobic conditions it forms the major link between carbohydrates, which are the almost exclusive respiratory substrate in higher plants (ap Rees, 1980a), and the tricarboxylic acid cycle. In so doing it also provides intermediates for biosynthesis (for example, phosphoenolpyruvate for phenylpropanoid synthesis) and small amounts of ATP and NADH. In anoxia, glycolysis is probably the sole source of ATP.

The significance of the oxidative pentose phosphate pathway must be the production of NADPH. This is the only product for which the complete pathway is needed; all intermediates required for biosynthesis can be made by a combination of glycolysis and the non-oxidative section of the pentose phosphate pathway. The only known fate of the NADPH formed by the pathway is its use in reductive biosynthesis. There is no reliable evidence that higher plants oxidise significant amounts of NADPH via the respiratory chain, either directly or via a transhydrogenase. It is possible that some NADPH is oxidised via soluble oxidases such as ascorbic acid oxidase, but there is no direct evidence that this is so (ap Rees, 1980a). Evidence that NADPH from the pathway is used in reductive biosynthesis is provided by the observation that many, but not all, such reductions are specific for NADPH. Correlations between the need for such NADPH and high activity of the pentose phosphate pathway have been noted (Pryke & ap Rees, 1976). More substantial evidence has been provided by demonstrating a direct transfer of hydrogen from glucose to the product of the reduction via NADPH formed in the pathway. This has been done for fat (Agrawal & Canvin, 1971) and lignin (Pryke & ap Rees, 1977) synthesis, and the rationale of the experiments is as follows. If [3-^{3}H]glucose is metabolised via the pentose phosphate pathway it will label NADPH via 6-phosphogluconate dehydrogenase. If [3-^{3}H]glucose enters glycolysis the label will be transferred to water during the isomerisation of the triose phosphates (Pryke & ap Rees, 1977). If the labelled NADPH is used in a reductive synthesis, then the product will become labelled. The extent to which ^{3}H moved from the glucose to the product because the ^{3}H was attached to the carbon atoms used to form the product is assessed independently by measuring the labelling of the product by [3-^{14}C]glucose.

The significance of the oxidative pentose phosphate pathway is that it produces reduced coenzyme which is immune from oxidation via the respiratory chain and thus is available for reductive biosyntheses. This does not mean that all reductive biosyntheses are driven by NADPH or that all NADPH is made via the pentose phosphate pathway. The facts that the oxidative reactions of the latter are distinct from glycolysis, and that glycolysis and the tricarboxylic acid cycle do not reduce NADP, mean that production of reducing power for biosynthesis and provision of substrate for the tricarboxylic acid cycle can proceed and be controlled independently.

In considering the importance of the location of pathways of carbohydrate oxidation in chloroplasts, it is important to bear in mind that in the chloroplasts these pathways almost certainly only operate in the dark. Operation in the light would short-circuit essential parts of the reductive pentose phosphate pathway. It is very probable that the prime role of the oxidative pentose phosphate pathway in chloroplasts is to prevent a drastic lowering of the NADPH:NADP$^+$ ratio when light-driven reduction of NADP$^+$ ceases in the dark. There is evidence that the pathway in the chloroplasts is inoperative in the light and is switched on in the dark, and that an appreciable proportion of the NADP in the chloroplast is reduced in the dark (Krause & Heber, 1976). Triose phosphate formed by the pathway in the chloroplast can be converted to 3-phosphoglycerate in the chloroplast. This could contribute to the maintenance of the level of ATP in the chloroplast in the dark. The significance of the glycolytic reactions in the chloroplast is probably twofold. They provide a direct route, independent of the need for NADPH, for ATP synthesis and for the conversion of chloroplast starch to a compound which can be exported to the cytosol. The activities of both pathways, and the degree of recycling through the pentose phosphate pathway, in isolated chloroplasts can be varied independently (Stitt & ap Rees, 1980a). It is likely that in chloroplasts in the leaf the relative activities of the two pathways, the degree of recycling, and whether triose phosphate or 3-phosphoglycerate is exported will vary according to the chloroplast's needs for NADPH and ATP, and according to the rate of starch breakdown.

Not enough is known about proplastids to reveal all the roles of glycolysis and the oxidative pentose phosphate pathway in these organelles. It is almost certain that one such role is the production of NADPH for those reductive biosyntheses which occur in proplastids; these include steps involved in the synthesis of fat and the assimilation of nitrate. The extent to which intermediates of the pathways in the proplastids provide carbon for the above syntheses is not known. A related problem which requires answering is what substrate is used by proplastids and how it enters these organelles.

I thank Dr F. D. Macdonald for his helpful criticisms.

References

Agrawal, P. K. & Canvin, D. T. (1971). The pentose phosphate pathway in relation to fat synthesis in the developing castor oil seed. *Plant Physiology*, **47**, 672–5.

ap Rees, T. (1974). Pathways of carbohydrate breakdown in higher plants. In *MTP International Review of Science, Biochemistry*, series 1, vol. XI, *Plant Biochemistry*, ed. D. H. Northcote, pp. 89–127. London: Butterworth.

ap Rees, T. (1977). Conservation of carbohydrate by the non-photosynthetic cells of higher plants. *Symposia of the Society for Experimental Biology*, **31**, 7–32.

ap Rees, T. (1980a). Assessment of the contributions of metabolic pathways to plant respiration. In *The Biochemistry of Plants*, vol. II, *Metabolism and Respiration*, ed. D. D. Davies, pp. 1–29. London & New York: Academic Press.

ap Rees, T. (1980b). Integration of pathways of synthesis and degradation of hexose phosphates. In *The Biochemistry of Plants*, vol. III, *Carbo-*

hydrates: Structure and Function, ed. J. Priess, pp. 1–42. London & New York: Academic Press.

ap Rees, T. & Beevers, H. (1960). Pathways of glucose dissimilation in carrot slices. *Plant Physiology*, **35**, 830–8.

ap Rees, T., Blanch, E., Graham, D. & Davies, D. D. (1965). Recycling in the pentose phosphate pathway: comparison of C_6/C_1 ratios measured with glucose-C^{14} and fructose-C^{14}. *Plant Physiology*, **40**, 910–13.

ap Rees, T., Fuller, W. A. & Green, J. H. (1981). Extremely high activities of phosphoenolpyruvate carboxylase in thermogenic tissues of Araceae. *Planta, Berlin*, **152**, 79–86.

Beevers, H. (1961). *Respiratory Metabolism in Plants*. Evanston: Row, Peterson.

Clark, M. G., Williams, J. F. & Blackmore, P. F. (1974). Exchange reactions in metabolism. *Catalysis Reviews. Science and Engineering*, **9**, 35–75.

Crawford, R. M. M. (1967). Alcohol dehydrogenase activity in relation to flooding tolerance in roots. *Journal of Experimental Botany*, **18**, 458–64.

Crawford, R. M. M. (1978). Metabolic adaptations to anoxia. In *Plant Life in Anaerobic Environments*, ed. D. D. Hook & R. M. M. Crawford, pp. 119–36. Ann Arbor, Michigan: Ann Arbor Science.

Davies, D. D. (1979). The central role of phosphoenolpyruvate in plant metabolism. *Annual Review of Plant Physiology*, **30**, 131–58.

Davies, D. D., Giovanelli, J. & ap Rees, T. (1964). *Plant Biochemistry*. Oxford: Blackwell Scientific.

Dennis, D. T. & Miernyk, J. A. (1982). Compartmentation of nonphotosynthetic carbohydrate metabolism. *Annual Review of Plant Physiology*, **33**, 27–50.

Dixon, W. L. & ap Rees, T. (1980). Identification of the regulatory steps in glycolysis in potato tubers. *Phytochemistry*, **19**, 1297–301.

Emes, M. J. & Fowler, M. W. (1979). Intracellular interactions between the pathways of carbohydrate oxidation and nitrate assimilation in plant roots. *Planta, Berlin*, **145**, 287–92.

Harbron, S., Foyer, C. & Walker, D. A. (1981). The purification and properties of sucrose-phosphate synthetase from spinach leaves: the involvement of this enzyme and fructose bisphosphatase in the regulation of sucrose biosynthesis. *Archives of Biochemistry and Biophysics*, **212**, 237–46.

Krause, G. H. & Heber, U. (1976). Energetics of intact chloroplasts. In *Topics in Photosynthesis*, vol. 1, ed. J. Barber, pp. 171–214. Amsterdam: Elsevier.

Miernyk, J. A. & Dennis, D. T. (1982). Isoenzymes of the glycolytic enzymes in endosperm of developing castor oil seeds. *Plant Physiology*, **69**, 825–8.

Murphy, D. J. & Leech, R. M. (1977). Lipid biosynthesis from [^{14}C]bicarbonate, [2-^{14}C]pyruvate and [1-^{14}C]acetate during photosynthesis by isolated spinach chloroplasts. *FEBS Letters*, **77**, 164–8.

Murphy, D. J. & Leech, R. M. (1978). The pathway of [^{14}C]bicarbonate incorporation into lipids in isolated photosynthesising spinach chloroplasts. *FEBS Letters*, **88**, 192–6.

Nishimura, M. & Beevers, H. (1979). Subcellular distribution of gluconeogenetic enzymes in germinating castor bean endosperm. *Plant Physiology*, **64**, 31–7.

Pryke, J. A. & ap Rees, T. (1976). Activity of the pentose phosphate pathway during lignification. *Planta, Berlin*, **132**, 279–84.

Pryke, J. A. & ap Rees, T. (1977). The pentose phosphate pathway as a source of NADPH for lignin synthesis. *Phytochemistry*, **16**, 557–60.

Simcox, P. D. & Dennis, D. T. (1978). 6-Phosphogluconate dehydrogenase isoenzymes from the developing endosperm of *Ricinus communis* L. *Plant Physiology*, **62**, 287–90.

Simcox, P. D., Reid, E. E., Canvin, D. T. & Dennis, D. T. (1977). Enzymes of the glycolytic and pentose phosphate pathways in proplastids from the developing endosperm of *Ricinus communis* L. *Plant Physiology*, **59**, 1128–32.

Smith, A. M. & ap Rees, T. (1979*a*). Effects of anaerobiosis on carbohydrate oxidation by roots of *Pisum sativum*. *Phytochemistry*, **18**, 1453–8.

Smith, A. M. & ap Rees, T. (1979*b*). Pathways of carbohydrate fermentation in roots of marsh plants. *Planta, Berlin*, **146**, 327–34.

Stitt, M. & ap Rees, T. (1978). Pathways of carbohydrate oxidation in leaves of *Pisum sativum* and *Triticum aestivum*. *Phytochemistry*, **17**, 1251–6.

Stitt, M. & ap Rees, T. (1979). Capacities of pea chloroplasts to catalyse the oxidative pentose phosphate pathway and glycolysis. *Phytochemistry*, **18**, 1905–11.

Stitt, M. & ap Rees, T. (1980*a*). Carbohydrate breakdown by chloroplasts of *Pisum sativum*. *Biochimica et Biophysica Acta*, **627**, 131–43.

Stitt, M. & ap Rees, T. (1980*b*). Estimation of the activity of the oxidative pentose phosphate pathway in pea chloroplasts. *Phytochemistry*, **19**, 1583–5.

Stitt, M., Bulpin, P. V. & ap Rees, T. (1978). Pathway of starch breakdown in photosynthetic tissues of *Pisum sativum*. *Biochimica et Biophysica Acta*, **544**, 200–14.

Streeter, J. G. & Thompson, J. F. (1972). Anaerobic accumulation of γ-aminobutyric acid and alanine in radish leaves (*Raphanus sativus* L.). *Plant Physiology*, **49**, 572–8.

Turner, J. F. & Turner, D. H. (1980). The regulation of glycolysis and the pentose phosphate pathway. In *The Biochemistry of Plants*, vol. II, *Metabolism and Respiration*, ed. D. D. Davies, pp. 279–316. London & New York: Academic Press.

Williams, J. F. (1980). A critical examination of the evidence for the reactions of the pentose phosphate pathway in animal tissues. *Trends in Biochemical Sciences*, **5**, 315–20.

Yamada, M., Usami, Q. & Nakajima, K. (1974). Long chain fatty acid synthesis in developing castor bean seeds. *Plant and Cell Physiology*, **15**, 49–58.

IAN M. MØLLER & JOHN M. PALMER

10 Regulation of the tricarboxylic acid cycle and organic acid metabolism

The tricarboxylic acid (TCA) cycle, also known as the citric acid cycle or the Krebs cycle, was first proposed more than 40 years ago by Sir Hans Krebs as a means for the oxidation of pyruvate. Biochemical studies showed that the enzymes responsible for the operation of the cycle were located in the mitochondrion. This demonstration was first achieved in animal tissues and later in plants. All the enzymes responsible for the catalysis of the cycle are located inside the inner mitochondrial membrane; some are attached to the inner surface of the cristae membrane while others are soluble in the matrix space. For the purpose of this review pyruvate dehydrogenase will be considered to be part of the TCA cycle. One full turn of the cycle will cause the conversion of pyruvate into carbon dioxide and reducing equivalents:

$$CH_3COCOO^- + 4NAD^+ + FAD^+ + 3H_2O$$
$$\rightarrow 3CO_2 + 4NADH + FADH_2 + 4H^+. \quad (1)$$

The reducing equivalents are subsequently oxidised by molecular oxygen via the respiratory chain to yield water plus ATP. It is of interest to note in the above stoichiometric equation that there are more hydrogens in the reducing equivalent pool (ten) than there were in the original pyruvate (four). The 'extra' hydrogens come from water molecules which were either added to hydrolyse CoA complexes as in citrate synthase and succinyl CoA synthetase, or added across a double bond as in the fumarase reaction.

Provided no intermediate is removed from the cycle, it can be maintained by only a catalytic amount of oxaloacetate which condenses with acetyl CoA to form citrate and is then regenerated at the end of the cycle. It is generally accepted that the purpose of the TCA cycle is to provide reducing equivalents for oxidation by the respiratory chain. The key reaction in this process is the dehydrogenation of the carbon substrate. The direct dehydrogenation of acetate is not possible under physiological conditions (Krebs, 1981). Dehydrogenation can occur, however, from a substrate where it is possible to extract a single hydrogen from each of two adjacent carbon atoms, e.g.

$$\text{HOOC—CH}_2\text{—CH}_2\text{—COOH} + \text{FAD}^+ \rightarrow$$
(succinate) $\text{HOOC—CH}=\text{COOH} + \text{FADH}_2.$ (2)
(fumarate)

In order to oxidise acetate completely using this method it is necessary to attach the acetate to some carbon acceptor, which is oxaloacetate in the case of the TCA cycle. To ensure efficient operation of the cycle it is desirable to have the acceptor regenerated at the end of the cycle.

Substrates for oxidation by the cycle come from the breakdown of high molecular weight compounds either as the consequence of mobilisation of storage materials such as starch or lipids or from the continuous turnover of cellular constituents such as enzymes, membranes or nucleic acids. Carbohydrates are degraded to pyruvate and enter the mitochondrion to be oxidised by the TCA cycle. The breakdown of fatty acids from lipids will give acetyl CoA which can be oxidised by the TCA cycle. However, in oil-rich seeds acetyl CoA is converted to succinate via the glyoxylate cycle located in the glyoxysomes. Proteins are converted into their constituent amino acids and these can enter the TCA cycle either as pyruvate and acetyl CoA or via α-ketoglutarate, succinyl CoA, fumarate and oxaloacetate.

Although the function of the TCA cycle is normally associated with catabolic processes, it is also important in supplying intermediates for the biosynthesis of cellular components. Thus, succinyl CoA is the precursor for cytochromes and chlorophylls and the keto-acids are involved in the synthesis of amino acids. Some of the cycle intermediates, notably malate and citrate, also accumulate in large amounts under some conditions and this appears to serve a physiological role; crassulacean acid metabolism, C_4 metabolism, cation uptake into roots, opening and closing of stomata and the ripening of fruit are important examples.

The TCA cycle thus forms a cross-road for catabolic and anabolic reactions in the plant cell and the regulation of its activity is of central importance. The subject has been reviewed recently by Wiskich (1980); therefore most of the references used in this chapter will be either review articles or relatively recent research papers.

Regulation of the TCA cycle
Regulation of the individual enzymes
   It is impossible to discuss the overall regulation of the complete cycle without a detailed knowledge of how each individual enzyme is regulated. In this section we discuss the properties of the enzymes and compare and contrast them with their counterparts in mammalian tissues.

Pyruvate dehydrogenase. Pyruvate dehydrogenase converts pyruvate to acetyl CoA, a reaction which in practice is irreversible. As discussed above, pyruvate and acetyl CoA both form important entry points into the TCA cycle and are also involved in biosynthetic reactions. In animal tissues where large amounts of TCA cycle intermediates are required for biosynthesis, pyruvate can also be converted into oxaloacetate (anaplerotic reaction) since under these conditions oxaloacetate would not be regenerated via the cycle. Furthermore, the activity of pyruvate dehydrogenase is dependent on NAD^+ and NADH. Clearly, a finely tuned regulation of pyruvate dehydrogenase is required to adjust to all of these factors. The complex pattern of regulation has been unravelled in detail in experiments on enzyme isolated from animal tissues and this regulation is shown in Fig. 1. The enzymes and the effectors which have so far been found to be functional in the plant enzyme system are underlined.

The enzyme has an inactive, phosphorylated form and an active, dephosphorylated form. The active form catalyses the reaction shown in Fig. 1. Acetyl CoA and NADH are competitive inhibitors for CoA and NAD^+, respectively (Rubin, Zahler & Randall, 1978).

The conversion of the active to the inactive form occurs by a MgATP-dependent phosphorylation mediated by a pyruvate dehydrogenase kinase which is part of the enzyme complex. Such a deactivation has been observed

Fig. 1. Mechanism of regulation of the pyruvate dehydrogenase complex from animal mitochondria. The figure is based mainly upon Williamson & Cooper (1980) and Wiskich (1980). Underlining means that the effector or enzyme has been found to be active in the plant enzyme complex. ' + ' denotes activator; ' − ' denotes inhibitor.

$$Pyruvate + CoA + NAD^+ \rightarrow acetyl\ CoA + CO_2 + NADH$$

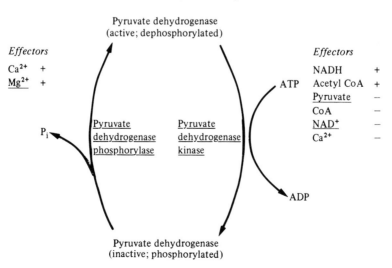

in the enzyme complex isolated from plant tissues (Rubin & Randall, 1977; Rao & Randall, 1980). The activation by dephosphorylation is catalysed by a phosphatase which appears to be absent in partly purified enzyme from plant tissues (Rubin & Randall, 1977), but activation can be observed when measuring activity in intact plant mitochondria (Rao & Randall, 1980). So it seems that the phosphatase is also present in the plant enzyme complex but easily lost or inactivated during purification.

As indicated in Fig. 1 the presence of substrates of the enzyme will inhibit the kinase and so the complex is kept active. Conversely, when products accumulate the kinase is activated, the phosphatase inactivated and the enzyme will be converted into the inactive form. Finally, it should be noted that the effect of calcium ions is similar to that of substrates in that it keeps the enzyme in its active form.

Citrate synthase. Citrate synthase causes a condensation of acetyl CoA and oxaloacetate to form citrate. In animal tissues the enzyme is inhibited by succinyl CoA, ATP, NADH and fatty acyl CoA. It is not known whether the inhibition by the latter three compounds has any physiological significance (Lehninger, 1975); however, the ATP effect appears to be artificial since MgATP, the predominant form of ATP in mitochondria, does not inhibit the enzyme (Williamson & Cooper, 1980). Succinyl CoA is a competitive inhibitor with respect to acetyl CoA and this feedback inhibition by a later intermediate in the cycle provides a link with α-ketoglutarate dehydrogenase which is also inhibited by succinyl CoA (see below). In animal tissues it appears to be the availability of the two substrates, acetyl CoA and oxaloacetate, which forms the most important regulatory mechanism.

The concentration of oxaloacetate is very dependent on the [NADH]/[NAD$^+$] ratio. Since this ratio also affects the dehydrogenases in the TCA cycle strongly its influence will be discussed below under the overall regulation of the activity of the cycle.

Citrate synthase has been isolated from several different plant tissues and is inhibited by ATP (competitive with respect to acetyl CoA, non-competitive with respect to oxaloacetate: Barbareschi *et al.*, 1974; Iredale, 1979). However, this inhibition is only 50% at 5 mM ATP and may not be of physiological significance. The plant enzyme differs from the mammalian enzyme by being insensitive to NADH (Iredale, 1979; Wiskich, 1980).

Aconitase. Aconitase catalyses the interconversion of citrate, aconitate and isocitrate and does not seem to be the site of any major regulation in animal tissues. The equilibrium constant favours the formation of citrate.

Isocitrate dehydrogenase. Isocitrate dehydrogenase can be either NAD^+- or $NADP^+$-specific. Both enzymes are found in animal as well as in plant mitochondria. The NAD^+-linked form is predominant. This enzyme is allosterically activated by AMP in microorganisms or ADP in mammalian systems and inhibited by ATP and high $[NADH]/[NAD^+]$ ratios. The plant enzyme has been relatively intensively studied, but shows no response to adenine nucleotides. However, it is activated by citrate and isocitrate and shows a strong dependence on the $[NADH]/[NAD^+]$ ratio, being 50% inhibited when the ratio $[NADH]/[NADH]+[NAD^+]$ is 0.1 (Duggleby & Dennis, 1970a, b).

α-Ketoglutarate dehydrogenase. α-Ketoglutarate dehydrogenase is an enzyme complex which is very similar to pyruvate dehydrogenase, but does not appear to be regulated by phosphorylation (Wedding & Black, 1971; Smith, Bryla & Williamson, 1974; Lehninger, 1975). In animals it is inhibited by high ratios of [succinyl CoA]/[CoA] and $[NADH]/[NAD^+]$ and this inhibition is relieved by increased concentrations of α-ketoglutarate. However, since high ratios of [succinyl CoA]/[CoA] and $[NADH]/[NAD^+]$ also inhibit isocitrate dehydrogenase the α-ketoglutarate necessary to overcome this inhibition would have to be supplied by some other pathway. This would, for example, be possible by the action of aspartate transferase (oxaloacetate + glutamate → aspartate + α-ketoglutarate) or glutamate dehydrogenase (L-glutamate + $NAD(P)^+ + H_2O \rightarrow$ α-ketoglutarate + $NH_3 + NAD(P)H + H^+$). Both of these reactions are central in the amino acid metabolism and the reader is referred to a review by Miflin & Lea (1977) for details on amino acid metabolism in plants. The plant α-ketoglutarate dehydrogenase is strongly activated by AMP and to a much lesser extent by ADP and ATP (Wedding & Black, 1971). We have not found any reports on the effect of NADH on the enzyme from plant tissues.

Succinyl CoA synthetase. Succinyl CoA synthetase converts succinyl CoA into succinate and produces ATP in the process. This 'substrate-level phosphorylation' differs from the reaction in animal tissues where GTP is the nucleotide involved (Palmer & Wedding, 1966).

Succinate dehydrogenase. Succinate dehydrogenase is a membrane-bound enzyme which donates its reducing equivalents directly to a flavoprotein in the respiratory chain (see Palmer, this volume). The isolated enzyme from both animal and plant sources is activated by ATP, ADP, reduced ubiquinone, acid pH, NADH, substrates (succinate and fumarate) and other treatments.

This activation is accompanied by the release of tightly bound oxaloacetate (Singer *et al.*, 1973; Oestreicher, Hogue & Singer, 1973). Whether the removal of bound oxaloacetate is the only cause of activation is not known. It is well known that succinate dehydrogenase needs to be activated to obtain maximal rates of succinate oxidation by isolated plant mitochondria. The pattern of this activation is similar to that of the isolated enzyme.

It is not clear whether the deactivation/activation of succinate dehydrogenase has a regulatory role. Succinate is normally oxidised faster by plant mitochondria than other cycle intermediates (Bowman, Ikuma & Stein, 1976) so if the substrate is supplied solely by the succinyl CoA synthetase there seems to be little need for control. However, in fat-containing seeds, where fatty acids are broken down, large amounts of succinate are produced by the glyoxysomes and under those conditions modulation could be useful.

Fumarase. No results have, to our knowledge, been published on the regulatory properties of fumarase isolated from plant tissues.

Malate dehydrogenase and the regulation of malate oxidation. Malate dehydrogenase (NAD^+-linked) catalyses the reaction:

$$\text{Malate} + NAD^+ \rightarrow \text{oxaloacetate} + NADH + H^+. \tag{3}$$

In animal tissues this enzyme does not appear to be the site of regulation whereas the purified plant enzyme has been reported to be strongly inhibited by ATP (Asahi & Nishimura, 1973). Two points speak against this effect having physiological importance. Firstly, the inhibition was competitive with respect to malate so that high malate concentrations would almost abolish any effect. Secondly, the activity of malate dehydrogenase is so high (see later) that even an 80% inhibition would hardly affect the function of the enzyme.

The pH optimum of malate dehydrogenase is 9.5 in the forward direction and 7.0 in the reverse (Asahi & Nishimura, 1973). The most important feature of malate dehydrogenase is that its equilibrium constant, K_{eq}, is extremely low (i.e. favours the substrates):

$$K_{eq} = [\text{OAA}][\text{NADH}]/[\text{MAL}][\text{NAD}^+] = 2.3 \times 10^{-5} \tag{4}$$

at pH 7.2 (where OAA is oxaloacetate, mal is malate and square brackets indicate concentration). This can be transformed to:

$$[\text{NADH}]/[\text{NAD}^+] = 2.3 \times 10^{-5} \times [\text{MAL}]/[\text{OAA}], \tag{5}$$

which means that it requires about 45 000 times more malate than oxaloacetate in the matrix to achieve a 50% reduction of the NAD^+ ($[\text{NADH}]/[\text{NAD}^+] = 1$). The level of NADH may determine the rate of respiration (see Palmer, this volume) and is, as we have already seen, involved in the regulation of a number of other TCA cycle enzymes. This means that the presence of even low levels of oxaloacetate can have quite a strong inhibitory effect on respiration (see later). As a consequence, animal mitochondria cannot oxidise malate unless oxaloacetate is removed either by condensation with acetyl CoA or by transamination with glutamate. Plant mitochondria, on the other hand, contain in their matrix an NAD^+-linked malic enzyme which converts malate into pyruvate:

$$\text{Malate} + NAD^+ \rightarrow \text{pyruvate} + NADH + CO_2. \tag{6}$$

The pH optimum of this enzyme is 6.8–7.0 (Macrae, 1971). The presence of malic enzyme allows plant mitochondria to oxidise malate as shown in Fig. 2 without the input of any other substrate (Palmer, 1976). The concerted action of malic enzyme and pyruvate dehydrogenase forms acetyl CoA which can then be used to remove malate. Thus, it takes two malate molecules to give one citrate which after one turn of the cycle regenerates one molecule of malate. This mechanism allows the interconversion of all TCA cycle intermediates and that means that the input of one intermediate can be used to supplement the pool of any other intermediate. This could be of major importance in tissues where large pools of organic acids are turned over, and will be discussed in the last section of this article.

Overall regulation of the TCA cycle

We have now discussed the regulation of the individual enzymes of the TCA cycle in plant tissues and the results are summarised in Fig. 3. Apart from the supply of substrate to the cycle, two parameters are important in the overall regulation of the cycle: (1) the energy charge or the level of adenine nucleotides, and (2) the reduction level of the pyridine nucleotides ($[NADH]/[NAD^+]$).

The energy charge. The energy charge, or the phosphorylation level, of the adenine nucleotides is given by:

$$\tfrac{1}{2}(2 \times [ATP] + [ADP])/([ATP] + [ADP] + [AMP]), \tag{7}$$

Fig. 2. The concerted action of malate dehydrogenase and malic enzyme.

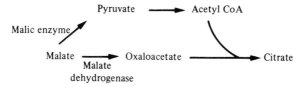

Net reaction: 2 malate + 3 NAD^+ → citrate + 3 NADH + 2 CO_2

and eventually → malate + 5 NADH + 1 $FADH_2$ + 1 ATP + 4 CO_2

and is the subject of the article by Pradet & Raymond in this volume. It has been postulated to be of central importance in the regulation of cellular metabolism. In animal mitochondria the activity of, for example, isocitrate dehydrogenase is dependent on the energy charge as the enzyme is activated by ADP and inhibited by ATP. Since the pyruvate dehydrogenase and citrate synthase are inhibited by ATP the TCA cycle in animal mitochondria will be turned off when the energy charge rises. Williamson & Cooper (1980) have, however, argued against the physiological significance of these observations. Firstly, the ATP inhibition of citrate synthase is probably not valid since the predominant form of ATP in mitochondria, MgATP, is not inhibitory. Secondly, changes in extramitochondrial [ATP]/[ADP] ratios within the physiologically relevant range only affect intramitochondrial ATP/ADP ratios marginally.

Before discussing the importance of energy charge in plant mitochondria two general points should be made regarding energy charge and adenine nucleotides. Firstly, it is important to recognise that experiments in which

Fig. 3. The TCA cycle in plant mitochondria and its regulation. 1, Pyruvate dehydrogenase; 2, citrate synthase; 3, aconitase; 4, isocitrate dehydrogenase; 5, α-ketoglutarate dehydrogenase; 6, succinyl CoA synthetase; 7, succinate dehydrogenase; 8, fumarase; 9, malate dehydrogenase. Red. UQ, reduced ubiquinone.

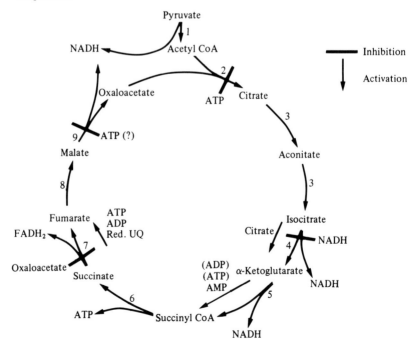

only ATP is added are carried out at an energy charge of 1.0, a value which is rarely or never observed in living tissues. Secondly, when only one of the adenine nucleotides affects an enzyme the concept of energy charge is less useful. In this case the concentration of that nucleotide is the relevant parameter and this will be governed by the concentration of the adenine nucleotides both in the cytoplasm and in the mitochondrion and by the properties of three enzymes: (1) adenylate kinase, which is located in the inner-membrane space and catalyses the interconversion of the adenine nucleotides $(2ADP \rightleftharpoons ATP + AMP)$; (2) the ADP/ATP translocator, which catalyses the exchange of ADP entering the matrix for ATP leaving it (Vignais, 1976); (3) an ADP transporter found in plant mitochondria that causes a net uptake of ADP (Abou-Khalil & Hanson, 1979a, b).

The only TCA cycle enzyme in plants to be markedly affected by adenine nucleotides appears to be pyruvate dehydrogenase, which is inactivated by MgATP (Fig. 1). In this respect plants differ from animals where the activity of isocitrate dehydrogenase is partly controlled by the energy charge. The physiological significance of the ATP inhibition of pyruvate dehydrogenase is difficult to evaluate since the enzyme complex is affected by so many factors (Fig. 1). It appears entirely possible, however, that a large change in the concentration of one of the other effectors could counteract the inhibition by ATP. For example, a low [NADH]/[NAD$^+$] ratio will favour the active form of the enzyme complex.

Plant mitochondria often contain a non-phosphorylating, alternative electron transport pathway (see Palmer, this volume). Its role has been suggested to be to allow a turnover of the TCA cycle under conditions of high energy charge when the normal respiratory chain is turned off (Palmer, 1976). A prerequisite for this theory is clearly that the TCA cycle itself is unaffected by a high energy charge.

The activity of the alternative pathway under high energy charge conditions will keep the [NADH]/[NAD$^+$] ratio relatively low, thereby counteracting the inhibitory effect of the high energy charge on pyruvate dehydrogenase. This is consistent with the suggested role of the alternative oxidase.

In summary, it is not obvious that either the energy charge or the level of adenine nucleotides has anything but a modifying influence on the turnover of the TCA cycle in plant tissues.

The reduction level of the pyridine nucleotides. In animal mitochondria the reduction level of the pyridine nucleotides is suggested to be the most important factor in the regulation of flow through the TCA cycle (Williamson & Cooper, 1980). On the basis of available evidence we think the same is true in plant mitochondria: the activities of all the four dehydrogenases in the cycle are dependent on the concentrations of NADH and NAD^+ (Fig. 3). We postulate that the $[NADH]/[NAD^+]$ ratio in plant mitochondria is determined primarily by the equilibrium position of malate dehydrogenase.

Malate dehydrogenase is much more active than other enzymes in the TCA cycle (Table 1; Bowman *et al.*, 1976) and will therefore always be near or at equilibrium. This means that relatively high oxaloacetate levels will give low NADH levels and vice versa due to the very low equilibrium constant of malate dehydrogenase (see eqn (5)). In support of this postulate can be mentioned that addition of oxaloacetate to mitochondria will inhibit not only the oxidation of malate, but also the oxidation of pyruvate, citrate or α-ketoglutarate measured as the consumption of oxygen. The inhibition of respiration is in fact much stronger with the latter three substrates, but in all cases transient (Brunton & Palmer, 1973). Recovery of respiration coincides with the removal of oxaloacetate. These results can be explained as follows. The addition of oxaloacetate causes malate dehydrogenase to act in reverse, the concentration of NADH decreases and respiration slows up. The NADH produced by the oxidation of substrates (by malic enzyme in the case of malate) will be used in this reversal of the malate dehydrogenase and not passed on to the respiratory chain until the oxaloacetate concentration has been lowered to a point where the NADH concentration has recovered.

Other observations on the effects of rotenone, NAD^+ and respiratory state support the postulate that the equilibrium of malate dehydrogenase determines the concentration of NADH in the matrix and therefore the rate of respiration (Brunton & Palmer, 1973; Tobin *et al.*, 1980; Palmer, Schwitzguébel & Møller, 1982).

Table 1. *Activities of TCA cycle enzymes in Jerusalem artichoke mitochondria*

Enzyme	Activity (nmol min^{-1} (mg protein)$^{-1}$)
Isocitrate dehydrogenase, NAD^+-linked	330
Succinate dehydrogenase	250
Fumarase (reverse reaction)	170
Malate dehydrogenase (forward reaction)	2500
Malate dehydrogenase (reverse reaction)	20000
Malic enzyme, NAD^+-linked (forward reaction)	140

Activities were measured at pH 7.2 in mitochondria disrupted with a detergent.

Cytoplasmic–mitochondrial interactions

In the previous section the regulation of the TCA cycle was discussed in isolation from the rest of the cell. It was not considered how events in the cytoplasm such as changes in pH, reduction levels of pyridine nucleotides or substrate concentration could affect the functioning of the TCA cycle. These points will be treated below.

Transport systems in mitochondria

The outer membrane of the mitochondrion presents no barrier to small molecules (below 10 000 daltons) while the inner membrane is selectively permeable and contains carriers or exchangers for a large number of metabolic intermediates and inorganic ions. Since this is treated elsewhere (see Moore and Earnshaw & Cooke, this volume) it suffices to point out here that the availability of the TCA cycle intermediates in the matrix will be a function of their concentration in the cytoplasm. As a number of the carriers catalyse an exchange, the concentration of an intermediate transported by such a system will also depend on the concentration of the counter-transported species in the matrix itself. The pH in the cytoplasm will play a role by affecting the state of ionisation of compounds with a pK within the physiological range as only one species of a given molecule is normally transported (e.g. ADP^{3-} for ADP: Vignais, 1976).

The effect of pH

D. D. Davies has proposed that two cytoplasmic enzymes with different pH optima – one carboxylating, one decarboxylating – can function as a pH-stat keeping the cytoplasmic pH within a relatively narrow range (Smith & Raven, 1979; Davies, this volume). If one looks at the pH optima of the enzymes of the TCA cycle and the external NAD(P)H dehydrogenases (Fig. 4) it is obvious that with two exceptions they all fall between pH 6.8 and pH 7.8. Both of the exceptions, citrate synthase and malate dehydrogenase (forward reaction), are very active and also show good activity within the above range (Barbareschi *et al.*, 1974; Palmer, 1980).

So there appears to be good opportunity for the cell to make major shifts in the relative rates of the cycle enzymes by small alterations in the pH of the cytoplasm. But is a change in pH transmitted across the inner mitochondrial membrane? According to the Mitchell hypothesis the mitochondrial inner membrane is relatively proton impermeable and this has been confirmed by experiments on plant mitochondria. However, an adjustment of matrix pH will take place over several minutes at neutral pH (Moore & Wilson, 1978). To illustrate this adjustment one can use the pH-dependent characteristics of malate oxidation by intact plant mitochondria. When the pH of the medium is 6.6 pyruvate is produced, while oxaloacetate is the only product at pH 8.0 (Palmer, Cowley & Al-Sané, 1978; Tobin *et al.*, 1980). In other words malic enzyme is most active at an external pH of 6.6 whereas it is inactive at pH 8.0. Since this is perfectly in keeping with the pH optimum of the isolated enzyme (see above) it shows that the external pH is transmitted to the matrix space in the physiologically important range pH 6.6–8.0.

It is important to keep in mind that pH in the bulk of the solution can be quite different from that found in the layer immediately adjacent to biological membranes. Due to negative charges on the membranes protons will be attracted, particularly if there are few other cations in the solution. The result is a lower local pH. The pH optimum of a membrane-bound enzyme will, therefore, have to be interpreted with caution. Furthermore, the meaning of pH is far from clear when one looks at the mitochondrial matrix. Due to the small volume, the matrix of one mitochondrion filled with solution would contain 30 protons at pH 7.0 and three at pH 8.0. Since the matrix contains

Fig. 4. The pH optima for TCA cycle enzymes and related enzymes in plant tissues. The data are from Wiskich (1980) and Møller & Palmer (1981).

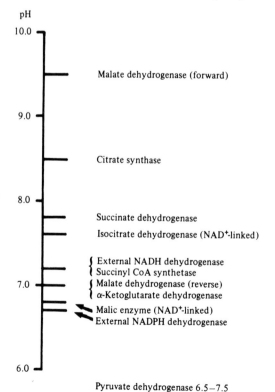

Pyruvate dehydrogenase 6.5–7.5

a very high concentration of proteins a large proportion of the water will be tied up as water of hydration (Srere, 1981) and this would serve further to destroy the traditional picture of pH in a dilute solution of buffer.

Reduction level of pyridine nucleotides in the cytoplasm

We have already seen how important the intramitochondrial ratio of [NADH] and [NAD$^+$] is in the regulation of the TCA cycle. Under conditions where biosynthesis is predominant, reducing equivalents produced in the mitochondria will be needed in the cytoplasm. Likewise, a major breakdown of lipids or carbohydrates in the glyoxysomes or in the cytoplasm will produce NAD(P)H which will, to some extent, be oxidised by the mitochondria. The reduction level of the pyridine nucleotides in the two compartments will be a decisive factor in determining which reactions will be favoured. This will be influenced by the relative concentrations of metabolites, the rates of the dehydrogenases and the pH. It is important to realise that the [NADH]/[NAD$^+$], [NADPH]/[NADP$^+$] and [ATP]/[ADP] ratios are linked and do not vary independently of each other (see Davies, this volume). In rat liver cells the [NADH]/[NAD$^+$] ratio is 0.15 in the cytoplasm and 0.6 in the mitochondria. However, if one looks at the concentration of free nucleotides only, these ratios are much lower (0.0004 and 0.1, respectively), reflecting the preferential binding of NADH to the dehydrogenases particularly in the cytoplasm (Akerboom, van der Meer & Tager, 1979). For plants such data are not available.

Until recently it was thought that the inner membrane of plant mitochondria was freely permeable to oxaloacetate, which would have the effect of equalising the reduction levels of the pyridine nucleotides between cytoplasm and mitochondria. However, it now appears that oxaloacetate movements are mediated by a transporter (Day & Wiskich, 1981; A. L. Moore, personal communication).

Plant mitochondria have the ability to oxidise NAD(P)H in the cytoplasm probably via two dehydrogenases located on the outer surface of the inner membrane (Møller & Palmer, 1981). Since it has been shown that the NADH dehydrogenase is Ca^{2+}-dependent (Møller, Johnston & Palmer, 1981) the possibility of a major involvement of calcium ions in the energy metabolism of plant cells is brought into focus (see Earnshaw & Cooke, this volume). Clearly, this external dehydrogenase will have an important effect on the way the cytoplasmic [NADH]/[NAD$^+$] ratio is regulated and we can, therefore, expect plant and mammalian cells to show marked differences in their regulation of pyridine nucleotide reduction levels.

Regulation of respiration by movement of coenzymes across the
inner mitochondrial membrane

The inner membrane of mammalian mitochondria is not permeable to coenzymes such as NAD^+ or thiamine pyrophosphate and no carriers exist for either. It has been assumed that the situation is the same in plant mitochondria, but recent evidence indicates that this may not be correct. NAD^+ and thiamine pyrophosphate both stimulate malate oxidation by isolated mitochondria quite strongly and the characteristics of this stimulation appear at present to suggest that the compounds cross the membrane (Tobin *et al.*, 1980; Palmer *et al.*, 1982; D. A. Moss & J. M. Palmer, unpublished). A mechanism for the net uptake of NAD^+ (or thiamine pyrophosphate) would clearly have to be very carefully regulated and this may be an area of future development in our understanding of the interaction between mitochondria and cytoplasm in plant cells.

Organic acid metabolism and the involvement of the mitochondria

We have now moved from the individual enzyme via the organelle and the whole cell up to the whole-plant level of complexity, and in this final section we will treat physiological conditions where the turnover of large pools of organic acids plays a metabolic role. Important examples are: (1) C_4 metabolism (Rathnam, 1978), (2) crassulacean acid metabolism (Osmond, 1978), (3) ion uptake into roots (Collins & Reilly, 1968), (4) fruit ripening (Beevers, Stiller & Butt, 1966) and (5) opening and closing of stomata (Raschke, 1975). In this review we have to limit ourselves to a cursory treatment of the first of these phenomena in order to assess the direct involvement of the mitochondria and cellular respiration.

The involvement of the mitochondria in C_4 metabolism

For a detailed review of the physiological characteristics of C_4 plants the reader is referred to Rathnam (1978). The main events are as follows: (1) carbon dioxide condenses with phosphoenolpyruvate in the mesophyll cells to form malate; (2) malate (or aspartate) is transported to the bundle sheath cells where it is decarboxylated to form a C_3 compound and carbon dioxide; (3) the released carbon dioxide is refixed by the Calvin cycle while the C_3 compound is transported back to the mesophyll cells.

It is clear that under conditions of active photosynthesis the size of fluxes of organic acids between the two cell types will be large although the pool size may stay relatively low depending on the position of the rate-limiting step. The decarboxylation in the bundle sheath cells is carried out by three different enzymes depending on the plant species and C_4 plants can be grouped according to which enzyme is predominant. The three enzymes are phosphoenolpyruvate carboxykinase, $NADP^+$-malic enzyme and NAD^+-malic enzyme. The last is a mitochondrial enzyme as already discussed. It is found in all three types of C_4 plants while the two former enzymes are located in the chloroplasts and are restricted to plants within their group. The direct involvement of mitochondria and an effect on cellular respiration is, therefore, clear in the NAD^+-malic enzyme C_4 plants. This involvement will be considered in a little more detail.

With aspartate as the starting compound the scheme shown in Fig. 5 has been proposed to cause a decarboxylation (Rathnam, 1978). Provided aspartate aminotransferase is present and there is a pool of α-ketoglutarate this scheme should work very well to produce pyruvate and carbon dioxide. The interesting points here are the fate of the pyruvate and the control of the malic enzyme. As indicated earlier a C_3 compound will have to be returned to the mesophyll cells so a mechanism must exist to avoid the oxidation of the pyruvate. This could be a high $[NADH]/[NAD^+]$ ratio due to restricted electron transport caused by a high energy charge (see Palmer and Pradet & Raymond, this volume). It could also be due to high acetyl CoA levels as a result of oxaloacetate inhibition of succinate dehydrogenase which would increase the concentration of succinyl CoA and inhibit citrate synthase (see earlier). Although this is speculation it is possible, and if pyruvate dehydrogenase inhibition were due to high acetyl CoA levels this would activate malic enzyme at the same time (Fig. 5).

Malic enzyme from several plant tissues is strongly activated by fructose-1,6-diphosphate (FDP: Davies & Patil, 1975; Chapman & Hatch, 1977). In intact mitochondria FDP prevents the loss of C_4 carboxylating capacity caused by incubation at 30 °C (Chapman & Hatch, 1977). Such effects are particularly interesting as they provide a link between glycolysis and respiration in both C_3 and C_4 plants and allow one to make predictions about overall regulation of plant respiration. The observation of an effect of FDP on intact mitochondria implies that the mitochondrial inner membrane has a receptor

Fig. 5. Decarboxylation in mitochondria of bundle sheath cells in C_4 plants of the NAD^+-malic enzyme type. I, aspartate aminotransferase; II, NAD^+-linked malate dehydrogenase; III, NAD^+-linked malic enzyme.

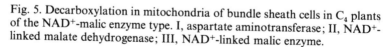

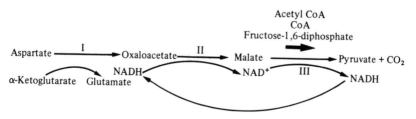

on its outer surface specific for FDP, a molecule which is not thought to be part of mitochondrial metabolism. This would be a very interesting control mechanism akin to that of hormones.

The previous paragraph is highly speculative but illustrates how our understanding of the respiration of plant mitochondria is becoming increasingly integrated into overall cellular metabolism.

I.M.M. acknowledges receipt of a NATO Science Fellowship and grants from the Danish and Swedish Natural Science Research Councils.

References

Abou-Khalil, S. & Hanson, J. B. (1979*a*). Energy-linked adenosine diphosphate accumulation by corn mitochondria. I. General characteristics and effect of inhibitors. *Plant Physiology*, **64**, 276–80.

Abou-Khalil, S. & Hanson, J. B. (1979*b*). Energy-linked adenosine diphosphate accumulation by corn mitochondria. II. Phosphate and divalent cation requirement. *Plant Physiology*, **64**, 281–4.

Akerboom, T. P. M., van der Meer, R. & Tager, J. M. (1979). Techniques for the investigation of intracellular compartmentation. In *Techniques in Metabolic Research* B205, pp. 1–33. Amsterdam: Elsevier/North-Holland.

Asahi, T. & Nishimura, M. (1973). Regulatory function of malate dehydrogenase isoenzymes in the cotyledons of mung bean. *Journal of Biochemistry, Tokyo*, **73**, 217–25.

Barbareschi, D., Longo, G. P., Servettaz, O., Zulian, T. & Longo, C. P. (1974). Citrate synthetase in mitochondria and glyoxysomes of maize scutellum. *Plant Physiology*, **53**, 802–7.

Beevers, H., Stiller, M. L. & Butt, V. S. (1966). Metabolism of the organic acids. In *Plant Physiology. A Treatise*, vol. IVB, ed. F. C. Steward, pp. 119–262. New York & London: Academic Press.

Bowman, E. J., Ikuma, H. & Stein, H. J. (1976). Citric acid cycle activity in mitochondria isolated from mung bean hypocotyls. *Plant Physiology*, **58**, 426–32.

Brunton, C. J. & Palmer, J. M. (1973). Pathways for the oxidation of malate and reduced pyridine nucleotide by wheat mitochondria. *European Journal of Biochemistry*, **39**, 283–91.

Chapman, K. S. R. & Hatch, M. D. (1977). Regulation of mitochondrial NAD-malic enzyme involved in C_4 pathway photosynthesis. *Archives of Biochemistry and Biophysics*, **184**, 298–306.

Collins, J. C. & Reilly, E. J. (1968). Chemical composition of the exudate from excised maize roots. *Planta, Berlin*, **83**, 218–22.

Davies, D. D. & Patil, K. D. (1975). The control of NAD specific malic enzyme from cauliflower bud mitochondria by metabolites. *Planta, Berlin*, **126**, 197–211.

Day, D. A. & Wiskich, J. T. (1981). Glycine metabolism and oxaloacetate transport by pea leaf mitochondria. *Plant Physiology*, **68**, 425–9.

Duggleby, R. G. & Dennis, D. T. (1970*a*). Nicotinamide adenine dinucleotide-specific isocitrate dehydrogenase from a higher plant. The requirement for free and metal-complexed isocitrate. *Journal of Biological Chemistry*, **245**, 3745–50.

Duggleby, R. G. & Dennis, D. T. (1970*b*). Regulation of the nicotinamide adenine dinucleotide-specific isocitrate dehydrogenase from a higher plant. The effect of reduced nicotinamide adenine dinucleotide and mixtures of citrate and isocitrate. *Journal of Biological Chemistry*, **245**, 3751–4.

Iredale, S. E. (1979). Properties of citrate synthase from *Pisum sativum* mitochondria. *Phytochemistry*, **18**, 1057.

Krebs, H. (1981). The evolution of metabolic pathways. In *Molecular and Cellular Aspects of Microbial Evolution*, Symposium of the Society for General Microbiology 32, ed. M. J. Carlile, J. F. Collins & B. E. B. Moseley, pp. 215–28. Cambridge University Press.

Lehninger, A. L. (1975). *Biochemistry*, 2nd edn. New York: Worth Publishers.

Macrae, A. R. (1971). Isolation and properties of a 'malic' enzyme from cauliflower bud mitochondria. *Biochemical Journal*, **122**, 495–501.

Miflin, B. J. & Lea, P. J. (1977). Amino acid metabolism. *Annual Review of Plant Physiology*, **28**, 299–329.

Møller, I. M., Johnston, S. P. & Palmer, J. M. (1981). A specific role for Ca^{2+} in the oxidation of exogenous NADH by Jerusalem-artichoke (*Helianthus tuberosus*) mitochondria. *Biochemical Journal*, **194**, 487–95.

Møller, I. M. & Palmer, J. M. (1981). Properties of the oxidation of exogenous NADH and NADPH by plant mitochondria. Evidence against a phosphatase or a nicotinamide nucleotide transhydrogenase being responsible for NADPH oxidation. *Biochimica et Biophysica Acta*, **638**, 225–33.

Moore, A. L. & Wilson, S. B. (1978). An estimation of the proton conductance of the inner membrane of turnip (*Brassica napus* L.) mitochondria. *Planta, Berlin*, **141**, 297–302.

Oestreicher, G., Hogue, P. & Singer, T. P. (1973). Regulation of succinate dehydrogenase in higher plants. II. Activation by substrates, reduced coenzyme Q, nucleotides, and anions. *Plant Physiology*, **52**, 622–6.

Osmond, C. B. (1978). Crassulacean acid metabolism. A curiosity in context. *Annual Review of Plant Physiology*, **29**, 379–414.

Palmer, J. M. (1976). The organization and regulation of electron transport in plant mitochondria. *Annual Review of Plant Physiology*, **27**, 133–57.

Palmer, J. M. (1980). The reduction of nicotinamide adenine dinucleotide by mitochondria isolated from *Helianthus tuberosus*. *Journal of Experimental Botany*, **27**, 418–30.

Palmer, J. M., Cowley, R. C. & Al-Sané, N. A. (1978). The inhibition of malate oxidation by oxaloacetate in Jerusalem artichoke mitochondria. In *Plant Mitochondria*, ed. G. Ducet & C. Lance, pp. 117–24. Amsterdam: Elsevier/North-Holland.

Palmer, J. M., Schwitzguébel, J.-P. & Møller, I. M. (1982). Regulation of malate oxidation in plant mitochondria: effect of NAD^+, respiratory state and rotenone. *Biochemical Journal*, **208**, 703–11.

Palmer, J. M. & Wedding, R. T. (1966). Purification and properties of succinyl-CoA synthetase from Jerusalem artichoke mitochondria. *Biochimica et Biophysica Acta*, **113**, 167–74.

Rao, K. P. & Randall, D. D. (1980). Plant pyruvate dehydrogenase complex: inactivation and reactivation by phosphorylation and dephosphorylation. *Archives of Biochemistry and Biophysics*, **200**, 461–6.

Raschke, K. (1975). Stomatal action. *Annual Review of Plant Physiology*, **26**, 309–40.

Rathnam, C. K. M. (1978). C_4 photosynthesis: the path of carbon in bundle sheath cells. *Science Progress, Oxford*, **65**, 409–35.

Rubin, P. M. & Randall, D. D. (1977). Regulation of plant pyruvate dehydrogenase complex by phosphorylation. *Plant Physiology*, **60**, 34–9.

Rubin, P. M., Zahler, W. L. & Randall, D. D. (1978). Plant pyruvate dehydrogenase complex: analysis of the kinetic properties and metabolite regulation. *Archives of Biochemistry and Biophysics*, **188**, 70–7.

Singer, T. P., Oestreicher, G., Hogue, P., Contreiras, J. & Brandao, I. (1973). Regulation of succinate dehydrogenase in higher plants. I. Some general characteristics of the membrane-bound enzyme. *Plant Physiology*, **52**, 616–21.

Smith, C. M., Bryla, J. & Williamson, J. R. (1974). Regulation of mitochondrial α-ketoglutarate metabolism by product inhibition at α-ketoglutarate dehydrogenase. *Journal of Biological Chemistry*, **249**, 1497–505.

Smith, F. A. & Raven, J. A. (1979). Intracellular pH and its regulation. *Annual Review of Plant Physiology*, **30**, 289–311.

Srere, P. A. (1981). Protein crystals as a model for mitochondrial matrix proteins. *Trends in Biochemical Sciences*, **6**, 4–7.

Tobin, A., Djerdjour, B., Journet, E., Neuburger, M. & Douce, R. (1980). Effect of NAD^+ on malate oxidation in intact plant mitochondria. *Plant Physiology*, **66**, 225–9.

Vignais, P. V. (1976). Molecular and physiological aspects of adenine nucleotide transport in mitochondria. *Biochimica et Biophysica Acta*, **456**, 1–38.

Wedding, R. T. & Black, M. K. (1971). Nucleotide activation of cauliflower α-ketoglutarate dehydrogenase. *Journal of Biological Chemistry*, **246**, 1638–43.

Williamson, J. R. & Cooper, R. H. (1980). Regulation of the citric acid cycle in mammalian systems. *FEBS Letters*, **117** (Supplement), K73–K85.

Wiskich, J. T. (1980). Control of the Krebs cycle. In *The Biochemistry of Plants. A Comprehensive Treatise*, vol. 2, ed. P. K. Stumpf & E. E. Conn, pp. 243–78. New York & London: Academic Press.

J. M. PALMER

11 The operation and control of the respiratory chain

A continual supply of gaseous oxygen is essential for the survival of higher plants, indicating that cellular respiration is an important metabolic process in plants. During cellular respiration molecular oxygen acts as the terminal acceptor for electrons originating from a variety of carbon substrates. The electrons are not donated directly from the substrate to oxygen, their transfer being mediated by a series of redox components capable of mutual oxidation and reduction; these components are collectively referred to as the respiratory chain.

Most of our understanding concerning the operation of the respiratory chain comes from the study of isolated mitochondria rather than the study of intact cells or tissue. The respiratory chain oxidises the reduced pyridine nucleotide coenzymes (NAD(P)H), and succinate produced mainly by the operation of the enzymes of the tricarboxylic acid (TCA) cycle. The principal purpose of respiration is considered to be the release of potential energy, available in the reduced substrates, in a form that can be readily converted into ATP. However, it is possible to consider that the oxidation of carbon substrates and their conversion into desirable end-products (such as succinyl CoA for the biosynthesis of cytochromes or chlorophyll) may be a desirable end in itself. Indeed, under suitable conditions it may be of greater importance than the energy transducing process. This could be especially true in higher plants where photosynthesis may be considered the principal energy-transducing pathway, and respiration may play an important role in inter-converting substrates during processes such as crassulacean acid metabolism or C_4/C_3 photosynthesis.

The organisation and control of the central respiratory chain

The central core of the respiratory chain is characterised by a linear sequence of mutually oxidisable and reducible components which transfer electrons from reduced coenzymes to molecular oxygen. The investigation of the behaviour of the chain has been facilitated firstly because most components are firmly bound to the inner membrane of the mitochondrion. Also, the system is coupled to the synthesis of ATP at three specific regions and is sensitive to inhibition at three specific points by the classical inhibitors of respiration, namely rotenone, antimycin A and cyanide (as outlined in Fig. 1).

In the mammalian mitochondrion electrons can be donated to the respiratory chain by the oxidisable acids of the TCA cycle and a variety of other substrates such as β-hydroxybutyrate, fatty acids and glycerol phosphate. The only donors capable of reducing the respiratory chain in plant mitochondria are the coenzymes reduced by the oxidation of acids of the TCA cycle.

Flavoproteins

The first components of the respiratory chain which oxidise the coenzyme NADH or succinate are believed to be flavins associated with the NADH and succinate dehydrogenases. In the mammalian system these flavoproteins have been isolated and characterised; the NADH dehydrogenase contains flavin mononucleotide (FMN) while succinate dehydrogenase contains flavin adenine dinucleotide (FAD). The dehydrogenases have not yet been isolated from plant mitochondria; therefore we have no direct information concerning the nature of the flavins involved and our current understanding is derived by analogy with the mammalian system.

Recently it has been possible to distinguish between different flavoproteins

Fig. 1. The linear arrangement of redox components which constitute the normal respiratory chain common to both plants and animals. The redox components closely associated with the three sites of ATP synthesis are indicated, together with the principal sites of action of respiratory inhibitors. The cytochrome components are represented by the lower-case letters a, b and c. FAD, flavin adenine dinucleotide, FeS, iron–sulphur protein; FMN, flavin mononucleotide; UQ, ubiquinone.

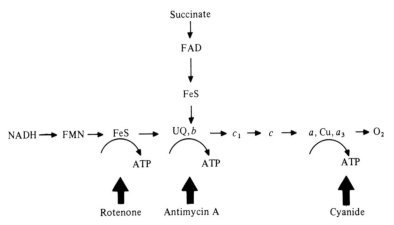

by measurement of their mid-point redox potentials (E_m) (Storey, 1980). This technique identifies five potential types of flavoprotein present in isolated plant mitochondria. The component with the lowest potential has an E_m (at pH 7.2) of -155 mV and is the only one that has a significant fluorescence; thus its assignment as a flavoprotein is relatively secure. The remaining four components are non-fluorescent and their assignment as flavins depends on the observation that they have an absorption peak at 464 nm which bleaches on reduction and that two electrons are involved in the complete reduction. This latter property distinguishes these components from iron–sulphur proteins, which have similar spectral properties but which only transfer one electron per mole. These four non-fluorescent flavins have E_m values in the region of -70 mV, $+20$ to $+40$ mV, $+110$ mV, and $+170$ to $+190$ mV respectively. The low-potential fluorescent flavin has been tentatively identified as the prosthetic group of lipoate dehydrogenase associated with the oxidation of α-keto-acids. It has not been possible to assign the other non-fluorescent flavins to any particular role in the respiratory chain although the component with an E_m of $+20$ to $+40$ mV does appear to undergo redox changes characteristic of its involvement in the cyanide-resistant oxidase (Storey, 1980). However, more data are required before it is certain this component is not an iron–sulphur protein.

Iron–sulphur proteins

The reduced flavin of the NADH dehydrogenase or succinate dehydrogenase is oxidised by iron–sulphur proteins. Current experimental evidence shows that in plant mitochondria there are at least three ferredoxin-type centres closely associated with the transfer of electrons from the flavin of the NADH dehydrogenase to ubiquinone (Cammack & Palmer, 1977). This is different from the mammalian respiratory chain where six iron–sulphur centres have been identified with this region (Ohnishi, 1972). The operation of the iron–sulphur centres in the NADH dehydrogenase is closely related to the operation of the first site of ATP synthesis and the inhibition of electron flux by rotenone (Palmer & Coleman, 1974). Three iron–sulphur centres also appear to be associated with the succinate dehydrogenase. Two of these, centres S_1 and S_2, are of the ferredoxin type and the third, centre S_3, contains four irons and four sulphurs (HiPIP type). The S_3 centre has a redox potential of $+80$ mV and works in close association with the ubiquinone. Of the two ferredoxin-type iron–sulphur proteins the S_1 centre has an E_m of -7 mV and appears to be involved in the transfer of electrons from succinate to the respiratory chain; the E_m of S_2 is -230 mV, and it is difficult to see how with such a negative redox potential it can play a role in the electron transfer sequence between succinate and ubiquinone.

In the mammalian system another iron–sulphur protein is involved in transferring electrons between cytochromes b and c_1 and is known as the Rieske iron–sulphur centre. Definitive data for the existence of such a centre in plant mitochondria are not yet available.

Ubiquinone

The reduced iron–sulphur centres of the succinate and NADH dehydrogenases are oxidised by ubiquinone, which is the next component in the electron transport sequence. The ubiquinone present in plant mitochondria is 2,3-methoxy-5-methyl-1,4-benzoquinone with a chain of ten isoprenyl derivatives substituted in position six; this is identical to that found in mammalian mitochondria. Ubiquinone is present in greater amounts than other carriers of the respiratory chain and is generally considered to be soluble in the lipid phase of the membrane, although recent studies have indicated that there may be a specific protein that can bind the ubiquinone (Yu, Yu & King, 1977) which may have the effect of restricting its mobility. Evidence from plant mitochondria suggests that the ubiquinone may not exist as a single homogenous pool; it appears to act as though it were organised in a series of sub-pools associated with different dehydrogenases (Palmer, 1979; Moore & Rich, 1980).

One of the interesting aspects of the role of ubiquinone in the respiratory chain is that in traditional schemes it oxidises iron–sulphur proteins and reduces cytochrome b; both of these carriers transfer only a single electron, whereas ubiquinone can accept either one or two electrons to form the ubisemiquinone radical or the fully reduced ubiquinol. The precise mechanism of electron transfer through this part of the respiratory chain is unresolved and made more difficult because of the complexity of cytochrome b. Analysis using electron paramagnetic resonance techniques shows that the ubisemiquinone free radical is formed during the respiratory process and is then further reduced to the ubiquinol by either accepting another electron from the respiratory chain (Rich & Moore, 1976) or interacting with another ubisemiquinone which would then rapidly disproportionate to give one ubiquinol and one ubiquinone.

Cytochromes

The terminal stages of electron transfer between ubiquinone and molecular oxygen are achieved by the cytochrome components of the respiratory chain. These cytochromes can be divided into three classes on the basis of the type of haem group present or the way the haem is bound to the protein. The a-type cytochromes contain haem A as the prosthetic group while cytochromes b and c both contain protoporphyrin IX as the haem. In the c-type cytochromes the haem is covalently bound to the protein while in b-type cytochromes the haem is bound only by electrostatic forces and can be easily removed by acidic pyridine. It is generally considered that the b-type cytochromes are closely associated with the oxidation of ubiquinol and that the c-type cytochromes are responsible for the transfer of electrons from the ubiquinone cytochrome b complex to cytochrome a,a_3, which is the only member of the respiratory chain that can interact directly with molecular oxygen.

The b-*type cytochromes.* The nature of cytochrome b in plants has always been considered complex. Spectral analysis has always shown the presence of at least three b-type cytochromes with α absorption bands at 556, 560 and 565 nm (at 25 °C). Early studies using animal mitochondria suggested that they contained only a single b-type cytochrome, although it has now been recognised that these also contain multiple forms of cytochrome b (Wikstrom, 1973). The significance of the presence of multiple types of cytochrome b remains unsolved. The mid-point redox potentials (E_m) of the different types of cytochrome b have been determined (Storey, 1980) and are b_{556}, $+88$ mV; b_{560}, $+79$ mV; and b_{565}, -76 mV (Lambowitz & Bonner, 1974). Kinetic analysis shows that these different b-type cytochromes behave in a very

Fig. 2. The ubiquinone cycle (after Mitchell, 1975, and Rich & Moore, 1976). This scheme shows the two-stage reduction of ubiquinone (UQ) to UQH_2, via the semiquinone radical UQH·, by electrons from the substrate and cytochrome b_{560} on the matrix side of the inner membrane. The oxidation takes place by a reverse process on the cytoplasmic side of the inner membrane. Antimycin A inhibits the ubiquinone cycle by preventing electron flow from b_{560} to the UQH·.

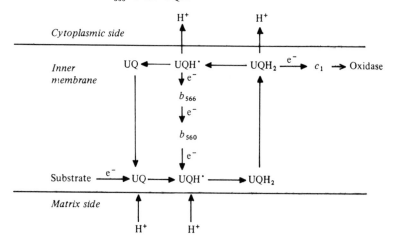

complex manner. The kinetic behaviour of cytochrome b_{560} is most easily understood; it is readily reduced by most substrates under anaerobic conditions and is rapidly re-oxidised when oxygen is made available. Cytochrome b_{556} is more complex in being extensively reduced only by succinate, in the absence of oxygen. Other substrates are less effective in reducing b_{556} than b_{560}, even though it has a slightly more positive E_m than cytochrome b_{560}. Cytochrome b_{556} reduced by succinate is re-oxidised but at a much slower rate than cytochrome b_{560}. The kinetic response of cytochrome b_{565} is very puzzling. It is not extensively reduced by either succinate or NADH under aerobic or anaerobic conditions in the absence of antimycin A. In the presence of antimycin A succinate will reduce most of cytochrome b_{565} under aerobic conditions. When the oxygen is exhausted, because of the activity of the antimycin-A-resistant alternative oxidase, the cytochrome b_{565} becomes oxidised. The oxidation of b_{565} seems to occur if electrons cannot flow away from cytochrome c to oxygen, possibly because the redox poise of the ubiquinone pool is disturbed.

One of the current developments in resolving the role played by the b-type cytochromes is the suggestion that they mediate electron flux between different levels of reduction of the ubiquinone as suggested in the operation of the ubiquinone cycle (Mitchell, 1975; Rich & Moore, 1976; see Fig. 2).

The c-type cytochromes. Spectral analysis shows that only one type of cytochrome c can be detected in plant mitochondria. In animal mitochondria, in contrast, there are two types, one of which is easily solubilised and has an α absorption peak at 550 nm while the other, cytochrome c_1, is more firmly bound to the membrane and has an α absorption peak at 554 nm. Plant mitochondria contain a small amount of soluble cytochrome c which appears to be spectrally and functionally the same as its counterpart in animal mitochondria. There is a large proportion of c-type cytochrome that cannot readily be removed from the membrane fraction, has an α absorption peak at 552 nm and may be considered the counterpart of cytochrome c_1 in animal systems. It has recently been possible to isolate a respiratory complex from plant mitochondria which contains cytochrome b and cytochrome c_{552} (Ducet & Diano, 1978), suggesting a close structural association between cytochrome c_{552} and the b-type cytochromes. Both cytochromes c_1 and c have E_m values of $+235$ mV and show kinetic responses consistent with a role in oxidising cytochrome b and reducing cytochrome oxidase.

The a-type cytochromes. The a-type cytochromes are the only components of the respiratory chain that can bind with and reduce molecular oxygen and thus they are ideally suited to form the final link in the respiratory chain. Plant mitochondria contain two a-type cytochromes which have slightly different spectral characteristics from their animal counterparts. The α absorption peak of the plant cytochrome is at 603 nm while that in animals is at 605 nm. The absorption seen at 603 nm is caused by the cytochrome a component; a spectral contribution attributable to cytochrome a_3 can only be measured in the Soret region of the absorption spectra, between 430 and 450 nm. The a-type cytochromes have two absorption peaks in this region, one strong peak at 438 nm and a weaker one at 445 nm. The absorption at 438 nm and 60% of the component at 445 nm is caused by cytochrome a while 40% of the peak at 445 nm is caused by the a_3 component (Storey, 1980).

The two a-type cytochromes can easily be distinguished by their mid-point redox potentials: cytochrome a has an E_m of $+190$ mV while cytochrome a_3 has an E_m of $+380$ mV. Kinetic and spectral studies carried out in the presence of potassium cyanide or carbon monoxide show that only the a_3 component interacts with these ligands, and thus it is possible to conclude that cytochrome a accepts electrons from cytochrome c and they are transferred to cytochrome a_3 which finally transfers them to molecular oxygen.

Characteristics of electron flow along the main respiratory chain

Electron flow along the main respiratory chain described in the previous section is characterised by the fact that it is inhibited by the classical inhibitors of respiration and that it is coupled to the synthesis of three moles of ATP.

Sensitivity to respiratory inhibitors

The three classical inhibitors of the central respiratory chain are rotenone, antimycin A and cyanide, and the central respiratory chains of animals, plants and microorganisms are similarly affected by these compounds.

Rotenone prevents the flow of electrons between the flavin of the NADH dehydrogenase and ubiquinone, probably by interacting with the iron–sulphur proteins. Thus it is a selective inhibitor of the oxidation of NADH. Antimycin A is an inhibitor that prevents electron transfer within the b–c_1 complex. Thus the addition of antimycin A causes redox components more reducing than cytochrome c to become reduced while allowing cytochromes c and a,a_3 to become oxidised. Antimycin A blocks the oxidation of all natural substrates by the main respiratory chain. Cyanide acts as an inhibitor by binding to cytochrome a_3 and preventing the terminal transfer of electrons from the respiratory chain to molecular oxygen.

Coupling of oxidation to the synthesis of ATP

The real physiological significance of the central respiratory chain is that it is coupled to the synthesis of ATP at three discrete positions. In the process of coupling it is necessary that the synthesis of ATP proceed simultaneously with the process of oxidation in order to obtain the maximal rate of respiration. The sequences of electron transport most closely associated with the synthesis of ATP correspond to the sites of inhibition by rotenone, antimycin A and cyanide. There is good experimental evidence to suggest that the central respiratory chain in plant mitochondria is as efficiently coupled to the synthesis of ATP as is the chain in any other type of mitochondria. The mechanism by which the coupling of ATP synthesis to the redox reactions is achieved is beyond the scope of this article and the reader is referred to Mitchell (1976) for further discussion.

Control of electron flux through the central respiratory chain

Little is really understood about the control of the rate of electron flux along the chain *in vivo*. The K_m of the cytochrome oxidase for oxygen is very low, in the region of 0.1 μM, and it is unlikely that the concentration of molecular oxygen is a limiting factor. There is some evidence to support the view that under certain conditions the supply of respiratory substrate is a limiting factor. This seems to be significant in regulating the rate of respiration in the developing spadix of *Arum maculatum* before the thermogenic burst of respiration. The most likely method of regulation is that exerted through the mechanism of respiratory control. In this system the lack of ADP, needed as an acceptor for inorganic phosphate during ATP synthesis, prevents the coupled rate of electron flux. Failure to synthesise ATP results in an increase in the magnitude of the proton-motive force, which becomes equal to the free energy available from the electron transport sequence responsible for driving the proton pump, and a thermodynamic equilibrium is rapidly established. Thus in the absence of ADP the rate of oxidation decreases and oxidisable substrate is conserved. This could be considered to be a potentially significant natural control mechanism but it would need a very high ratio of ATP to ADP in the cell and this could exert allosteric inhibition of certain key enzymes associated with carbon metabolism, such as phosphofructokinase or isocitrate dehydrogenase. It would also prevent oxidation by lack of substrate supply before respiratory control could be exerted.

Alternative electron transport sequences present in many plant mitochondria

The presence of alternative NADH dehydrogenase and terminal oxidase sequences of electron transport represents one of the unique differences between plant mitochondria and the mammalian system.

Cyanide-resistant alternative oxidase

The presence of an alternative terminal oxidase system that is resistant to inhibition by cyanide is a well-documented property of many plant mitochondria. It is absent in the mitochondria isolated from certain storage tubers such as potatoes and Jerusalem artichokes, but is a characteristic of the intact tubers of such tissue, and cyanide-resistant oxidation can be detected in mitochondria isolated from slices of such tissue after washing in a dilute salt solution for 24 hours. Therefore information necessary for the synthesis of the oxidase is present in this tissue but may not necessarily be expressed in the phenotype. It is possible that in tissue such as potato and Jerusalem artichoke the oxidase may be present in the intact tuber but is lost during the process of extraction of the mitochondria, possibly because of extensive degradation of phospholipid by lipase enzymes during the isolation of the organelles (see Laties, 1982).

Characteristics of the cyanide-resistant terminal oxidase

The most obvious characteristic of the oxidase is that it remains active in the presence of cyanide and antimycin A. Experiments by Bendall & Bonner (1971) clearly showed that the oxidase was different from cytochrome oxidase. The nature of the redox components responsible for the oxidase remain unresolved although there is a very large literature on the subject. A great increase in our ability to study the oxidase occurred when Schonbaum *et al.* (1971) showed that it was inhibited by benzhydroxamic acids. It was initially thought that these inhibitors specifically affected the alternative oxidase but it is now known that the benzhydroxamates can inhibit polyphenol oxidases and lipoxygenase. This lack of specificity makes it difficult to interpret results carried out on intact tissue such as slices or seeds, where it is often considered that inhibition of oxygen uptake by the benzhydroxamates demonstrates that the alternative pathway is involved.

The original idea that the oxidase was a flavoprotein was abandoned in favour of it being an iron–sulphur protein. This notion has now also dropped from favour and the problem remains unsolved. In recent years it has been possible to extract a quinol oxidase from *Arum* mitochondria which has sufficiently high capacity to be the oxidase, and is sensitive to benzhydroxamates (Huq & Palmer, 1978a). This fraction contains both a flavoprotein and copper, both of which could be involved in the oxidase activity, but positive proof awaits further experimentation, and the possibility that the oxidase may be a manifestation of the auto-oxidation of a quinone free radical has not yet been excluded.

Relationship between the alternative oxidase and the central respiratory chain

The cyanide-resistant oxidase is generally considered to be a branch of the central respiratory chain, accepting electrons from the level of ubiquinone. There is clear evidence to confirm that ubiquinone is the electron donor. Erecinska & Storey (1970) first obtained kinetic evidence that the cyanide-resistant oxidase accepted electrons from ubiquinone, and more recently Huq & Palmer (1978b) showed that extraction of quinones with pentane prevented electron flow through the cyanide-resistant oxidase, and that the activity of the oxidase could be regained by adding back an appropriate quinone. However, the behaviour of the ubiquinone pool is very complex and recent evidence shows that in mitochondria extracted from tissue not associated with thermogenesis, not all substrates are equally able to supply electrons to the cyanide-resistant oxidase. Tomlinson & Moreland (1975) and Huq & Palmer (1978b) showed that external NADH is readily oxidised by cytochrome oxidase but has only limited access to the alternative oxidase, while succinate could supply reducing equivalents to both oxidases. This is extremely difficult to understand if one assumes that both substrates supply electrons to a single pool of ubiquinone and that the oxidases then accept electrons from this homogenous pool of ubiquinol. Thus it is necessary to consider that the ubiquinone does not act as though it were a single pool but rather is organised as separate kinetic pools which prefer to interact with selected dehydrogenases and terminal oxidases (Palmer, 1979; Moore & Rich, 1980). The view of the oxidase activity as a simple branch from ubiquinone is therefore possibly an over-simplification of the situation. The nature of the different pools of ubiquinone is not clear; they could be either bound or soluble ubiquinone or could represent different oxidation states such as the ubisemiquinone or ubiquinol, depending on whether the process involves a single- or two-electron transfer process.

It is interesting to note that all mitochondria which contain an active cyanide-resistant oxidase also show a high degree of resistance to rotenone; in fact respiration in the presence of cyanide is unaffected by antimycin A or rotenone, suggesting that electrons can be transferred from the substrate to oxygen by a redox chain totally separate from the central chain already described. It is significant that when thin slices of storage tissue are washed to induce the cyanide-resistant oxidase there is a parallel increase in the level of rotenone-resistant oxidation. It is therefore possible that the cyanide-resistant oxidase represents the terminal step of a parallel, alternative electron transport chain as shown in Fig. 3. In this scheme it is recognised that electrons can cross over between the two systems at the level of ubiquinone, although it is not clear whether this occurs *in vivo* or whether it is an artefact that can be induced using isolated organelles in the presence of appropriate inhibitors.

It is interesting to speculate on the relationship of cyanide-resistant oxidation to the synthesis of ATP. It is generally held that the oxidation of ubiquinol by the cyanide-resistant oxidase does not result in the synthesis of ATP. The non-phosphorylating nature of the cyanide-resistant oxidation has important implications when considering the metabolic role of the alternative oxidation processes.

Control of oxidation via the alternative oxidase

Little is known about oxidation via the alternative oxidase and there are two theories currently advanced to explain factors that control the relative activities of the two oxidases.

Bahr & Bonner (1973*a*, *b*) have published what has become the classical

Fig. 3. Scheme (*a*) shows the conventional branched pathway (Bahr & Bonner, 1973*a*, *b*) in which electrons traverse the first site of oxidative phosphorylation and rotenone-sensitive iron–sulphur centres before reaching the branch-point. Scheme (*b*) shows two parallel chains, the dotted arrow between the two pools of quinone indicating a possible cross-over between the two redox systems. In this scheme electrons can reach oxygen via a totally non-phosphorylating pathway. The cytochrome components are represented by the lower-case letters *a*, *b* and *c*. FeS, iron–sulphur proteins; FMN, flavin mononucleotide; Fp, flavoprotein; UQ, ubiquinone.

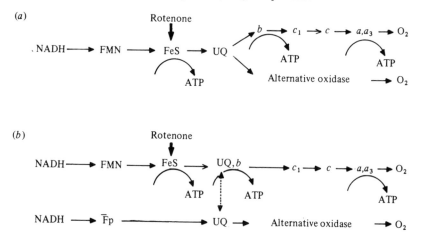

theory of control (Day, Arron & Laties, 1980). In these studies they titrated the respiratory rate with benzhydroxamates both in the presence of cyanide (g(i)) and in its absence (V_T). They then plotted g(i) against V_T and deduced that the cytochrome oxidase pathway was always operating at its maximal velocity whilst the alternative oxidase carried any excess electrons. This method also allowed them to determine the fraction of the alternative oxidase present that was actually engaged under any set of metabolic conditions (ρ). They concluded that ρ was numerically equal to the 'fractional reduction' of the compound that donates electrons to the alternative oxidase. It is apparent that Bahr & Bonner considered the 'fractional reduction' of the donor to the alternative oxidase as representing the redox poise of that component, and they assumed that the velocity of the electron flux through the alternative oxidase was directly related to the redox poise of this undefined donor species. It is interesting that linear plots for V_T against g(i) can be obtained, because it would be expected that as the benzhydroxamate caused inhibition of the alternative oxidase the redox poise of the donor would have become more negative and thus the fraction of the remaining functional oxidase involved in the respiratory process would increase as the benzhydroxamate concentration was raised. This would result in a departure of the value ρ from constancy.

De Troostembergh & Nyns (1978) have proposed an alternative model to explain the kinetics of electron flux through the two oxidase systems in yeast mitochondria. This scheme is based on a 'random' competition of the two oxidase systems for reducing equivalents originating from ubiquinol. De Troostembergh & Nyns (1978) claim that their model can explain the puzzling relationship between g(i) and V_T found by Bahr & Bonner and they interpret the concept of 'fractional reduction' of the donor for the alternative oxidase as being a limited fraction of the whole pool of the donor capable of working at any one time – which is clearly different from the redox poise of the complete pool. Data obtained in my own laboratory using *Arum* mitochondria show that as the rate of succinate oxidation is reduced, by adding malonate, the flux of electrons through both the cytochrome oxidase and the alternative oxidase is reduced. It seems, therefore, premature to consider that the mechanism which controls the relative activity of the oxidases is resolved.

Alternative systems for the oxidation of NADH

There appear to be two systems capable of oxidising NADH in addition to the normal rotenone-sensitive NADH dehydrogenase associated with the main respiratory chain. Both of these alternative NADH dehydrogenase systems are resistant to inhibition by rotenone. One is capable of oxidising exogenous NADH while the other can oxidise endogenous NADH.

Characteristics of the external NADH dehydrogenase

Isolated plant mitochondria have long been known to oxidise NADH added to the assay medium and it was originally considered that this was achieved by the NADH gaining access to the matrix space through damaged membranes. However, the use of ferricyanide as a non-penetrating acceptor (Palmer & Coleman, 1974) showed that the initial stages of oxidation of exogenous NADH occurred by an NADH dehydrogenase located outside the inner membrane. It is generally assumed that the dehydrogenase is bound to the outer surface of the inner membrane and donates electrons via a flavoprotein to ubiquinone. This pathway for the oxidation of external NADH is not sensitive to rotenone and is not coupled to ATP synthesis at the first site of oxidative phosphorylation.

Control of NADH oxidation via the external NADH dehydrogenase

It is often observed that the redox poise of the cytosolic and mitochondrial NADH is quite different, the NAD/NADH ratio in the cytosol being more reduced than that in the mitochondrial matrix. It is therefore reasonable to assume that some mechanism exists to regulate the relative rates of NADH oxidation from these two different metabolic compartments. Several observations have been made which allow us to speculate on potential control mechanisms for the rate of operation of the external NADH dehydrogenase. Firstly the ionic composition of the assay medium has always played a major role in influencing the contribution made by the external system. Recent data have helped to clarify the reasons for these ionic effects (Møller, Johnston & Palmer, 1981), for it has been shown that cations may play two roles. Firstly they act in a general way to neutralise the fixed negative charges associated with the membrane structure, i.e. lipids and proteins. The neutralisation and screening of these fixed charges allows the NADH, which bears a negative charge itself, to approach the active site of the dehydrogenase without experiencing excessive electrostatic repulsion, which is reflected in a change in the apparent K_m of the enzyme for NADH. The screening of the fixed membrane charges by cations also influences the apparent V_{max} for the oxidation of NADH, suggesting that the enzyme may undergo a conformational change to a more favourable state or that, because the oxidation is a multi-step process, the reduction in charge density allows a greater fluidity of the membrane components and thus increases the collision frequency resulting in an overall increase in the rate of oxidation. In addition to this rather general charge-screening effect there is a specific effect of calcium ions which promote the oxidation of external NADH, and in the absence of adequate calcium the beneficial influence of charge-screening cannot be observed. The dependency on calcium ions can be induced by adding the chelator EGTA to lower the calcium bound to the mitochondrial membrane. The addition of calcium in the presence of EGTA will restore NADH oxidation. However, if EGTA is removed, by washing, then the rate of NADH oxidation recovers without the need to add extra calcium and it appears that calcium is mobilised from within the mitochondrial structure to replace that removed by treatment with EGTA.

Another potential control mechanism is related to the observation that when external NADH and succinate are supplied simultaneously as substrates the final rate of oxygen consumption is less than that predicted if the rate of oxidation of each individual donor supplied alone were added together (Cowley & Palmer, 1980). Spectroscopic analysis shows that the oxidation of external NADH is strongly inhibited by the presence of succinate whilst the oxidation of succinate remains unaffected by the presence of NADH. Therefore the oxidation of succinate appears to turn off the oxidation of external NADH; this is a potentially important control mechanism and at present there is no rational explanation of how these competing electron fluxes interact with each other.

The physiological significance of the external pathway for NADH oxidation remains unknown, but it does seem to be a general feature of most isolated plant mitochondria.

Rotenone-resistant oxidation of internal NADH

In recent years it has been recognised that it is possible to oxidise endogenous NADH via a rotenone-resistant NADH dehydrogenase. This enzyme is distinct from the external NADH dehydrogenase in that it is not inhibited by EGTA, is not apparently dependent on calcium ions, and also operates in the absence of added pyridine nucleotide. Electron transport via this enzyme is not coupled to the synthesis of ATP at the first coupling site. One important aspect of research into plant mitochondria is to elucidate the mechanism which controls the electron flux through the two internal NADH dehydrogenases.

Control of electron flux through the two internal NADH dehydrogenase systems

There are relatively few papers that deal with the control of electron flux through the two internal NADH dehydrogenase systems and more information is necessary before we can be certain of the mechanisms which control the relative contribution that each of the NADH dehydrogenases makes to the overall respiratory process.

Sotthibandhu & Palmer (1975) observed that NAD^+-linked substrates could not be oxidised rapidly in the presence of a weak acid uncoupler unless an adenine nucleotide was present, and concluded that AMP appeared to act as a specific activator of the rotenone-sensitive NADH dehydrogenase. Other studies have shown that the behaviour of the internal NADH dehydrogenase system is peculiar when malate is used as the substrate. It has been generally observed that when malate is oxidised the rate of oxygen uptake is more strongly inhibited by rotenone than when other NAD$^+$-linked substrates are used. The rotenone-resistant rate of malate oxidation does, however, appear to recover after a reasonable period of time (Brunton & Palmer, 1973). This finding was initially interpreted as being due to some form of internal compartmentation in the matrix of the mitochondrion which causes the NADH produced by the malic enzyme to be oxidised by the rotenone-resistant dehydrogenase and the NADH produced by the malate dehydrogenase to be oxidised by the rotenone-sensitive NADH dehydrogenase (Brunton & Palmer, 1973). It seems, however, that this is a complex argument that is difficult to justify in reality and some simpler explanation would be preferable. One possibility could be that when malate is supplied as the electron donor the malate dehydrogenase plays a principal role in the processes of oxidation, although the NAD$^+$-linked malic enzyme is also involved. Because the malate dehydrogenase is a very active enzyme and the equilibrium constant favours the reverse reaction (i.e. conversion of oxaloacetate to malate) it is reasonable to assume that the malate dehydrogenase reaction is always at, or very near, equilibrium and as the reaction proceeds and the respiratory chain steadily oxidises the NADH and consequently the level of oxaloacetate will gradually build up and the redox poise of the pyridine nucleotide pool in equilibrium with the malate and oxaloacetate will become progressively oxidised. In this very oxidised state the concentration of NADH becomes insufficient to act as a donor to the rotenone-resistant dehydrogenase and only the rotenone-sensitive dehydrogenase is able to mediate electron flux and the respiration becomes sensitive to rotenone. When the inhibitor is present the activity of the malic enzyme continues to reduce the pool of pyridine nucleotide, but since the malate dehydrogenase is always in equilibrium the NADH level can only rise if the oxaloacetate level falls, and this represents the reaction taking place during the period before the recovery of rotenone-resistant oxidation of malate.

There is, therefore, a case for suggesting that one of the factors that regulates the activity of the NADH dehydrogenases is the concentration of NADH. It is possible that the rotenone-sensitive NADH dehydrogenase has a higher affinity for NADH than does the rotenone-resistant dehydrogenase and that oxaloacetate can act as a 'buffer' regulating the redox state of the mitochondrial pyridine nucleotide pool and dictates which electron transport sequence may be active in any metabolic state. Recent research tends to confirm this view. Tobin *et al.* (1980) and Palmer, Schwitzguebel & Møller (1982) have shown that rotenone causes a rapid decrease in the concentration of oxaloacetate which is consistent. Møller & Palmer (1982) have shown that the K_m values for the rotenone-sensitive and rotenone-resistant NADH dehydrogenases are 7 and 80 μM respectively, which is consistent with the proposed method of regulation. The fact that the level of oxaloacetate is very small when the rotenone-resistant NADH dehydrogenase is active means that the level falls below that necessary to operate the citrate synthase and therefore the complete TCA cycle cannot operate under these conditions. Thus the inhibitor-resistant NADH dehydrogenase can only sustain acid interconversion between succinate and malate or pyruvate and not the complete oxidation of pyruvate (Palmer & Møller, 1982).

Some of the work described in this article was carried out with the aid of research grants from the Science Research Council, The Royal Society, and the Central Research Fund, University of London.

References

Bahr, J. T. & Bonner, W. D. (1973*a*). Cyanide-insensitive respiration. I. The steady states of skunk cabbage spadix and bean hypocotyl mitochondria. *Journal of Biological Chemistry*, **248**, 3441–5.

Bahr, J. T. & Bonner, W. D. (1973*b*). Cyanide-insensitive respiration. II. Control of the alternative pathway. *Journal of Biological Chemistry*, **248**, 3446–50.

Bendall, D. S. & Bonner, W. D. (1971). Cyanide-insensitive respiration in plant mitochondria. *Plant Physiology*, **47**, 236–45.

Brunton, C. J. & Palmer, J. M. (1973). Pathways for the oxidation of malate and reduced pyridine nucleotide by wheat mitochondria. *European Journal of Biochemistry*, **39**, 283–91.

Cammack, R. & Palmer, J. M. (1977). Iron–sulphur centres in mitochondria from *Arum maculatum* spadix with very high rates of cyanide resistant respiration. *Biochemical Journal*, **166**, 347–55.

Cowley, R. C. & Palmer, J. M. (1980). The interaction between exogenous NADH oxidase and succinate oxidase in Jerusalem artichoke (*Helianthus tuberosus*) mitochondria. *Journal of Experimental Botany*, **31**, 199–207.

Day, D. A., Arron, G. P. & Laties, G. G. (1980). Nature and control of respiratory pathways in plants: the interaction of cyanide-resistant respiration with the cyanide-sensitive pathway. In *The Biochemistry of Plants*, vol. 2, *Metabolism and Respiration*, ed. D. D. Davies, pp. 198–237. New York & London: Academic Press.

De Troostembergh, J.-C. & Nyns, E.-J. (1978). Kinetics of the respiration of cyanide-insensitive mitochondria from the yeast *Saccharomycopsis lipolytica*. *European Journal of Biochemistry*, **85**, 423–32.

Ducet, G. & Diano, M. (1978). On the dissociation of the cytochrome *b–c* of potato mitochondria. *Plant Science Letters*, **11**, 217–26.

Erecinska, M. & Storey, B. T. (1970). The respiratory chain of plant mitochondria. VII. Kinetics of flavoprotein oxidation in skunk cabbage mitochondria. *Plant Physiology*, **46**, 618–24.

Huq, S. & Palmer, J. M. (1978*a*). Superoxide and hydrogen peroxide production in cyanide resistant *Arum maculatum* mitochondria. *Plant Science Letters*, **11**, 351–8.

Huq, S. & Palmer, J. M. (1978*b*). The involvement and possible role of ubiquinone in cyanide resistant respiration. In *Plant Mitochondria*, ed. G. Ducet & C. Lance, pp. 225–32. Amsterdam: Elsevier/North-Holland.

Lambowitz, A. M. & Bonner, W. D. (1974). The *b*-cytochromes of plant mitochondria. A spectrophotometric and potentiometric study. *Journal of Biological Chemistry*, **24**, 2428–40.

Laties, G. G. (1982). The cyanide-resistant, alternative path in higher plant respiration. *Annual Review of Plant Physiology*, **33**, 519–55.

Mitchell, P. (1975). Protonmotive redox mechanisms of the cytochrome *b–c* complex in the respiratory chain: protonmotive ubiquinone cycle. *Federated European Biochemical Societies Letters*, **56**, 1–6.

Mitchell, P. (1976). Vectorial chemistry and the molecular mechanics of chemiosmotic coupling: power transmission by proticity. *Biochemical Society Transactions*, **4**, 399–430.

Møller, I. M., Johnston, S. P. & Palmer, J. M. (1981). A specific role for Ca in the oxidation of exogenous NADH by Jerusalem artichoke mitochondria. *Biochemical Journal*, **194**, 487–95.

Møller, I. M. & Palmer, J. M. (1982). Direct evidence for the presence of a rotenone-resistant NADH dehydrogenase on the inner surface of the inner membrane of plant mitochondria. *Physiologia Plantarum*, **54**, 267–74.

Moore, A. L. & Rich, P. R. (1980). The bioenergetics of plant mitochondria. *Trends in Biochemical Sciences*, **5**, 284–8.

Ohnishi, T. (1972). Mechanism of electron transport and energy conservation in the site I region of the respiratory chain. *Biochimica et Biophysica Acta*, **301**, 105–28.

Palmer, J. M. (1979). The 'uniqueness' of plant mitochondria. *Biochemical Society Transactions*, **7**, 246–52.

Palmer, J. M. & Coleman, J. O. D. (1974). Multiple pathways of NADH oxidation in mitochondria. In *Horizons in Biochemistry & Biophysics*, vol. 1, ed. E. Quagliariello, F. Palmieri & T. P. Singer, pp. 220–60. Reading, Mass.: Addison-Wesley.

Palmer, J. M. & Møller, I. M. (1982). Regulation of NAD(P)H dehydrogenases in plant mitochondria. *Trends in Biochemical Sciences*, **7**, 258–61.

Palmer, J. M., Schwitzguebel, J.-P. & Møller, I. M. (1982). Regulation of malate oxidation in plant mitochondria. *Biochemical Journal*, **208**, 703–11.

Rich, P. R. & Moore, A. L. (1976). The involvement of the protonmotive ubiquinone cycle in the respiratory chain of higher plants and its relation to the branchpoint of the alternative pathway. *Federated European Biochemical Societies Letters*, **65**, 339–44.

Schonbaum, G. R., Bonner, W. D., Storey, B. T. & Bahr, J. T. (1971). Specific inhibition of the cyanide-insensitive respiratory pathway in plant mitochondria by hydroxamic acids. *Plant Physiology*, **47**, 124–8.

Sotthibandhu, R. S. & Palmer, J. M. (1975). The activation of non-phosphorylating electron transport by adenine nucleotides in Jerusalem artichoke (*Helianthus tuberosus*) mitochondria. *Biochemical Journal,* **152,** 637–45.

Storey, B. T. (1980). Electron transport and energy coupling in plant mitochondria. In *The Biochemistry of Plants,* vol. 2, *Metabolism and Respiration,* ed. D. D. Davies, pp. 125–7. New York & London: Academic Press.

Tobin, A., Djerdjour, B., Journet, E., Neuburger, M. & Douce, R. (1980). Effect of NAD^+ on malate oxidation in intact plant mitochondria. *Plant Physiology,* **66,** 225–9.

Tomlinson, P. F. & Moreland, D. E. (1975). Cyanide-resistant respiration of sweet potato mitochondria. *Plant Physiology,* **55,** 365–9.

Wikstrom, M. K. F. (1973). The different cytochrome *b* components in the respiratory chain in animal mitochondria and their role in electron transport and energy conservation. *Biochimica et Biophysica Acta,* **301,** 155–93.

Yu, C. A., Yu, L. & King, T. E. (1977). The existence of a ubiquinone binding protein in the reconstitutively active cytochrome *b–c* complex. *Biochemical and Biophysical Research Communications,* **78,** 259–65.

ANTHONY L. MOORE

12 Interactions between mitochondria and chloroplasts in higher plants

The manner in which photosynthetic events influence mitochondrial respiration and the degree to which mitochondrial metabolism is controlled by light have long been a problem. In particular controversy has surrounded the question of whether mitochondrial respiration continues in the light and if so to what extent. Observations have ranged from total inhibition to stimulation or to having no effect at all. The answer to this question does have considerable implications with respect to estimates of plant photosynthesis since net carbon dioxide fixation has to be corrected for losses of carbon dioxide due to mitochondrial and photorespiratory activity. Photorespiration is a light-dependent release of previously fixed carbon dioxide which occurs via a series of complex reactions that involve a close interaction between chloroplasts, mitochondria and peroxisomes.

Mitochondria house the enzymes of the tricarboxylic acid (TCA) cycle and are generally considered to be the 'powerhouse' of the cell, whereas chloroplasts are the sites at which light energy is utilised in the photosynthetic reduction of carbon dioxide to carbohydrates. The discovery that chloroplasts had the capability of synthesising ATP by photophosphorylation (Arnon, Allen & Whatley, 1954) made the participation of respiratory ATP in fulfilling cellular energy requirements in the light unnecessary. Energy production is not, however, the only function of mitochondria and certainly it is becoming increasingly evident that in plants mitochondria may play a more biosynthetic role than their mammalian counterparts. Carbon skeletons are required for biosynthetic purposes and this would require a turnover of the TCA cycle under conditions of high photosynthetic activity, thus suggesting that some degree of mitochondrial activity occurs in the light.

This article will attempt to consider the interactions between mitochondrial respiration and photosynthesis during illumination with respect to metabolite fluxes and control by adenine nucleotides. The role of mitochondrial respiration in photorespiration will also be considered. Obviously the interactions between mitochondria and chloroplasts are very complex, their full comprehension requiring a much more detailed description than is possible here, and the reader is referred to a number of excellent review articles (Heber, 1974, 1975; Krause & Heber, 1976; Schnarrenberger & Fock, 1976; Strotmann & Murakami, 1976; Graham & Chapman, 1979; Heber & Walker, 1979; Giersch, Heber & Krause, 1980*a*; Graham, 1980; Heber & Heldt, 1981) in which this topic has been considered.

Structural and functional similarities of mitochondria and chloroplasts

Mitochondria and chloroplasts share a number of similar properties. Both organelles possess an envelope system comprising two membranes which enclose a matrix or stroma consisting of soluble constituents such as enzymes and, in the case of chloroplasts, chlorophyll-containing vesicular membranes called thylakoids. Both organelles are involved in energy conservation, ATP synthesis being associated with the matrix side of the membranous system. The light reactions of photosynthesis such as primary quantum conversion, oxygen production, electron transport and photophosphorylation are associated with the internal membranes whereas the fixation and reduction of carbon dioxide occur in the matrix or stromal phase (Halliwell, 1978). Likewise electron transport in mitochondria is located in the inner membrane and the matrix contains the soluble enzymes of the TCA cycle (see Palmer, this volume). In mitochondria, oxygen serves as the terminal electron acceptor oxidising NADH of the TCA cycle, whereas in chloroplasts $NADP^+$ is the physiological electron acceptor and electrons are donated by water. Unlike the situation in animal tissues, the enzymes of β-oxidation are not normally found in plant mitochondria (Mazliak, 1973), being specifically associated with certain microbodies called glyoxysomes. Enzymes of the glycolytic and oxidative pentose phosphate pathway are generally regarded to be located principally in the cytosol, although the enzymes of the pentose phosphate pathway have been found in chloroplasts (Heber, Hallier & Hudson, 1967).

In the chloroplasts, carbon dioxide is fixed into organic products via a carboxylation reaction in which one molecule of ribulose bisphosphate condenses with one molecule of carbon dioxide, the reaction being catalysed by ribulose bisphosphate carboxylase. The reaction product, 3-phosphoglyceric acid, is then converted to triose phosphates by a series of reactions which are in essence the reverse of the glycolytic breakdown of carbohydrates which occurs in the cytosol. In this reaction 3-phosphoglyceric acid is phosphorylated to give diphosphorylglyceric acid which is then reduced to form glyceraldehyde-3-phosphate. Isomerisation of this compound yields dihydroxyacetone phosphate which then undergoes condensation with glyceraldehyde-3-phosphate. Sugar phosphate is the net product of photosynthesis although part of it undergoes a series of sugar interconversions resulting in the regeneration of the primary carbon dioxide acceptor molecule, ribulose bisphosphate. The Benson–Calvin cycle, as it is referred to (see Halliwell, 1978), requires the input of energy and reducing equivalents in the form of ATP and NADPH which are produced in the photochemical and electron transfer events of the light reactions. Only one of the products of photosynthesis, namely starch, is retained in the chloroplasts and then only temporarily; other products are immediately exported.

Mitochondria, in contrast to chloroplasts, are the sites of the terminal catabolism of carbohydrates, lipids and proteins and contain the enzymes of the TCA cycle. Full accounts of the TCA cycle and electron transport components can be found elsewhere in this volume as well as in all general biochemistry texts and will not be given here. In essence this cycle is similar to the Benson–Calvin cycle in so much as it is initiated by a condensation reaction in which acetyl CoA combines with oxaloacetate to give citrate which is then metabolised by a number of oxidation and decarboxylation reactions to regenerate oxaloacetate. The cycle provides precursors for the synthesis of sugars, fats and amino acids by interacting with other biosynthetic pathways. In the operation of the cycle oxidation reactions catalysed by pyruvate, isocitrate, 2-oxoglutarate and malate dehydrogenases are coupled to the reduction of NAD^+ to NADH (in the case of succinate dehydrogenase to the reduction of FAD to $FADH_2$). The re-oxidation of these nucleotides is via the electron transport chain, resulting in the production of ATP.

While both organelles are the sites of ATP synthesis, numerous energy-requiring biosynthetic reactions proceed outside these organelles, examples are the synthesis of proteins, fatty acids, polysaccharides and amino acids. Such a compartmentation of reactions necessitates the transfer of energy and metabolites from chloroplasts and mitochondria to the cytosol.

Metabolite transport in mitochondria and chloroplasts

The outer membrane of both organelles is unspecifically permeable to substances of low molecular weight such as nucleotides, amino acids and mannitol but not to large molecules such as polymeric sugars or proteins (Pfaff, Heldt & Klingenberg, 1969; Heldt & Sauer, 1970) (see Fig. 1). In contrast the inner membrane exhibits very specific permeability properties, being permeable only to carbon dioxide, oxygen and water and certain monocarboxylic acids in their protonated form (Heldt & Sauer, 1970; Klingenberg, 1970*a*, *b*). The inner membrane is therefore the functional barrier between the matrix or stroma and the cytosol. This impermeability of the inner membrane is overcome by the presence in each organelle of specific carriers for the transport of phosphate, dicarboxylates and adenine nucleotides (Klingenberg, 1970*a*, *b*; Heldt, 1976*a*, *b*; Wiskich, 1977, 1980; Heber & Heldt, 1981).

The specificity and activity of the translocators of either organelle vary somewhat (Fig. 1). For instance, in mitochondria, phosphate uptake is normally in exchange for hydroxyl ions and is highly specific, whereas in chloroplasts, 3-phosphoglycerate and dihydroxyacetone phosphate can compete with the uptake of phosphate (Heldt, 1976*a*, *b*). However, the uptake is normally by exchange and for each phosphate taken up another phosphate, phosphoglycerate or dihydroxyacetone phosphate is released from the stroma. Both organelles are capable of translocating dicarboxylates. In contrast to the mitochondrial carrier, however, the translocator in chloroplasts has a very wide specificity for various C_4 and C_5 dicarboxylates including glutamate and aspartate (Heldt, 1976*a*, *b*; Heber & Heldt, 1981), both of which are associated with the α-ketoglutarate carrier in mitochondria. Dicarboxylate transport was initially attributed to a single translocator, but the possibility is still open that transport might be due to different carriers with overlapping specificity. In addition to a dicarboxylate and α-ketoglutarate carrier, mitochondria possess a tricarboxylate carrier which transports citrate and the dicarboxylates malate and succinate (Palmieri *et al.*, 1972). Such a carrier is absent in chloroplasts.

Fig. 1. Schematic diagram of the transport characteristics of mitochondria and chloroplasts. A, anion; DHAP, dihydroxyacetone phosphate; PGA, 3-phosphoglycerate; P_i, inorganic phosphate.

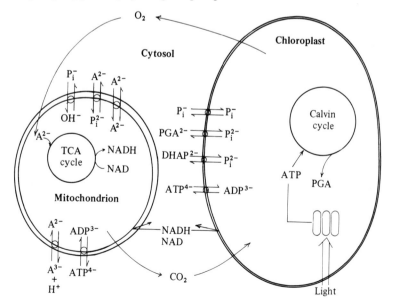

In both organelles there is a counter exchange of adenine nucleotides into the stroma or matrix mediated via the adenine nucleotide translocase (Heldt, 1969; Vignais *et al.*, 1976; Wiskich, 1977; Klingenberg, 1979). In mitochondria adenine nucleotide translocation enables the export of ATP generated by oxidative phosphorylation at the matrix side of the inner membrane to the cytosol. The specificity of the adenine nucleotide carrier is different for the two directions of transport. For transport into mitochondria, ADP is favoured more than ATP (Pfaff *et al.*, 1969), whereas the reverse is true for the outward transport of ATP. This asymmetrical exchange tends to generate a considerably higher ATP/ADP ratio outside the mitochondria than inside it, and appears to be regulated by the membrane potential. In contrast, the translocator has been less well studied in chloroplasts and its metabolic role is not fully understood (Heldt, 1969, 1976a, b). Although it proceeds by counter-exchange in a similar way to the mitochondrial translocator, it is highly specific for external ATP. It has been suggested that it may play a role in providing the chloroplast with ATP during the dark phase (Heldt, 1976a; Heber & Heldt, 1981).

From the preceding discussion it is apparent that the inner membrane of both organelles is impermeable to pyridine nucleotides. However, there is evidence that the transfer of pyridine nucleotides across the membrane may be effected indirectly via the metabolite shuttles. The dicarboxylate translocator can transport oxaloacetate and malate, and malate dehydrogenase, which catalyses the oxidation of malate to oxaloacetate with NAD^+ as the oxidant, is present in the cytosol as well as in the chloroplasts and mitochondria. Thus the malate/oxaloacetate shuttle may indirectly transfer reducing equivalents in the form of NADH between mitochondria, chloroplasts and the cytosol (Fig. 2) (Heber & Krause, 1972; Krause & Heber, 1976). Although this shuttle may exist *in vivo*, doubts have been raised as to the physiological concentration of oxaloacetate in the chloroplast (Heldt, 1976a), and it has been suggested that in chloroplasts at least oxaloacetate may well undergo transamination with glutamate leading to the formation of α-ketoglutarate and aspartate (Fig. 2) (Heber, 1975). Such a system has been demonstrated to occur in mammalian mitochondria (La Noue, Walajtys & Williamson, 1973).

Cytoplasmic NADPH is also required for biosynthetic purposes and it has been proposed that its provision may be facilitated indirectly by the chloroplast phosphate translocator. As suggested earlier, this translocator is capable of transferring both 3-phosphoglycerate and dihydroxyacetone phosphate. It is suggested that 3-phosphoglycerate is reduced in the chloroplast stroma to dihydroxyacetone phosphate at the expense of internal ATP and NADPH (Fig. 2). Dihydroxyacetone phosphate is exported into the cytosol in exchange for 3-phosphoglycerate where it is re-oxidised by a triose phosphate dehydrogenase concomitantly reducing NADP. Alternatively it is conceivable that re-oxidation to 3-phosphoglycerate may be effected via the glycolytic pathway yielding NADH and ATP. In this way NADPH and ATP are indirectly transferred from the chloroplast stroma to the cytosol. Heldt (1976a) has suggested, however, that the export of reducing equivalents in the form of NADPH takes priority over the supply of ATP and NADH to the cytosol.

In mitochondria, the shuttle systems appear to be more involved in the continued transfer of intermediates for both biosynthetic and oxidative purposes, the transfer of reducing equivalents taking a minor role. In this respect, it is important to remember that plant mitochondria possess an external NADH dehydrogenase which may facilitate the oxidation of cytosolic NADH via the respiratory chain (Douce, Mannella & Bonner, 1973; Day

Fig. 2. Schematic representation of metabolite transfer between chloroplasts, cytosol and mitochondria during illumination. Illumination produces a low cytosolic, high stromal 3-phosphoglycerate to dihydroxyacetone phosphate (PGA/DHAP) ratio resulting in indirect transport of NADH and ATP to the cytosol. NADH may either be transferred back to the chloroplast via the malate/oxaloacetate (Mal/OAA) shuttle or alternatively be re-oxidised by the mitochondrial respiratory chain. The size of the cytosolic ATP/ADP ratio will depend upon ATP utilisation such as in protein or sucrose synthesis. Lowered ATP/ADP ratios result in mitochondrial respiratory activity. Such activity will allow TCA cycle turnover, thus supplying carbon skeletons for biosynthetic purposes. Asp, aspartate; Glut, glutamate; α-Kg, α-ketoglutarate.

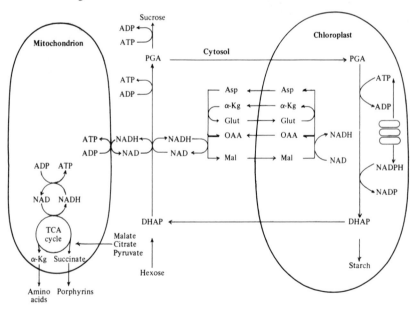

& Wiskich, 1974). Whether the operation of a mitochondrial malate/aspartate shuttle contributes to the oxidation of cytosolic NADH is uncertain. ATP is, of course, preferentially exported to the cytosol via the mitochondrial adenine nucleotide translocator.

Changes in metabolite levels in chloroplasts upon illumination

In vitro, the major products exported from the chloroplast stroma are dihydroxyacetone phosphate, 3-phosphoglycerate and glycolate. Since the former two metabolites are translocated via the phosphate carrier, its activity regulates their internal concentration (Heber, 1974; Heldt, 1976*a*; Heber & Walker, 1979; Heber & Heldt, 1981). Indeed, its main role is the transport of fixed carbon from the chloroplast stroma to the cytosol. The internal ratio of phosphoglycerate and dihydroxyacetone phosphate determines, in turn, the distribution of photosynthetic products in the light and the mobilisation of products such as starch in the dark. As dihydroxyacetone phosphate and phosphoglycerate also transfer reducing equivalents between the stroma and cytosol it is evident too that their ratios in these compartments determine the stromal and cytosolic redox states of the pyridine nucleotides. Thus concentration gradients are in effect maintained by the translocators (Heber, 1974; Krause & Heber, 1976; Heber & Walker, 1979). These in turn appear to be influenced by a proton gradient which is set up in the light across the chloroplast envelope (Heber, 1975; Heber & Heldt, 1981). Since protons are involved in pyridine nucleotide reactions, a proton gradient will result in differences in the redox state of the pyridine nucleotides on both sides of the chloroplast envelope.

In the light, protons are pumped from the stroma into the intrathylakoid space, which thus becomes acidic while the pH in the stroma increases. When this occurs phosphoglycerate is trapped in the alkaline stroma. This is because the phosphate translocator only transfers divalent anions (Lilley *et al.*, 1977) and at alkaline pH phosphoglycerate exists mainly as the trivalent anion. Thus alkaline pH causes the preferential export of dihydroxyacetone phosphate resulting in a low external, high internal ratio of phosphoglycerate to dihydroxyacetone phosphate ([PGA]/[DHAP]). Since dihydroxyacetone phosphate can be re-oxidised in the cytosol via the glycolytic pathway to yield ATP, low [PGA]/[DHAP] ratios are accompanied by high cytosolic [ATP]/[ADP] ratios. When the pH in the stroma decreases, there is a decrease in the [PGA]/[DHAP] ratio which is accompanied by an increase in the stromal [ATP]/[ADP] ratio.

For the continued operation of a [PGA]/[DHAP] shuttle *in vivo*, a mechanism to re-oxidise cytosolic NADH is required. This may be achieved either by the mitochondrial external NADH dehydrogenase or, alternatively, cytosolic NADH may be transferred back into the chloroplast by the malate/oxaloacetate shuttle. There is evidence to suggest that such a system transfers hydrogen across the chloroplast envelope (Heber & Krause, 1971) in both directions, depending on the physiological situation. Indeed, studies on intact leaf cells suggest that, in the light, the pyridine nucleotides are much more reduced in the chloroplasts than in the cytosol (Heber & Santarius, 1970).

Thus if the dicarboxylate and the phosphate translocator operate simultaneously it is not difficult to envisage a situation involving these two translocators in which a high [ATP]/[ADP] ratio outside exists side by side with a low [ATP]/[ADP] and a high [NADH]/[NAD$^+$] ratio inside the chloroplasts. This, in fact, appears to be the situation that exists in the leaf cell in the light (Heber, 1974). The driving force which maintains these gradients is the trans-envelope proton gradient which is developed in the light. This gradient is the result of an active extrusion of protons from the stroma into the external space. Whether this is due to an ATP-driven electrogenic pump actually located in the envelope is at present uncertain (Heber & Heldt, 1981). In addition to the trans-envelope gradient there is also a light-driven proton transport into the thylakoids. The activity of this pump is almost two orders of magnitude higher than observed rates of active proton extrusion from the stroma into the external medium (Heber & Heldt, 1981). Net proton extrusion from the stroma thus results in the pH of the stroma rising by almost 1 pH unit (Heldt et al., 1973).

There is now a considerable body of experimental evidence to support the hypothesis that upon illumination the level of phosphoglycerate is higher in the stroma than in the cytosol, whereas the reverse situation occurs with dihydroxyacetone phosphate (Lilley et al., 1977; Stitt, Wirtz & Heldt, 1980). This decreased cytosolic phosphoglycerate level and increased levels of triose phosphate produce a high [ATP]/[ADP] ratio in the cytosol (Heber & Santarius, 1970). Recent experiments suggest that although the [ATP]/[ADP] ratio in the chloroplast stroma does rise (to approximately 2–3), the ratio in the cytosol is very much higher (Giersch, Heber & Krause, 1980a; Hampp, Goller & Ziegler, 1982; Stitt, Lilley & Heldt, 1982). With respect to pyridine nucleotides it has been found that upon illumination a strong reduction occurs only in the stroma while in the cytosol no significant increase in the reduced to oxidised ratio can be detected (Giersch et al., 1980b).

The rise in ATP in the chloroplast stroma upon illumination is accompanied by changes in ADP and AMP, although the total adenylate contents remain constant (Stitt et al., 1982). In this respect it is of importance to consider the ratio [ATP]/[ADP][P$_i$], which is a measure of the free energy of ATP hydrolysis and is known as the phosphorylation potential (Chance &

Williams, 1956), since not only does it affect ATP-utilising reactions such as phosphoglycerate reduction but it may also be involved in the control of electron flow. Another term describing the energy potential of a system is the 'energy charge' (Atkinson, 1977), given by the ratio:

$$\frac{1}{2} \frac{[ADP]+2[ATP]}{[AMP]+[ADP]+[ATP]},$$

which is a measure of the extent to which the adenylate system is phosphorylated. It can assume any value between one (all in the form of ATP) and zero (all adenylates as AMP). Recent determinations (Kobayashi *et al.*, 1979; Giersch *et al.*, 1980a; Hampp *et al.*, 1982) suggest that both the phosphorylation potential and the energy charge rapidly increase in the chloroplast stroma upon illumination. In particular the energy charge rose to 0.8, which is known to be a region of optimal regulation and interaction of energy-yielding and energy-conserving reactions. Similarly stromal phosphorylation potential increased by a factor of 2–3 upon illumination, although it must be noted that the phosphorylation potentials measured were very much lower than previously reported (3×10^4 M^{-1} compared with 0.3×10^3 M^{-1}: Giersch *et al.*, 1980a). Few measurements have been made on the cytosolic energy charge or phosphorylation potential but recent determinations by Hampp *et al.* (1982) show that there is a fast transfer of photosynthetic phosphorylation power to the cytosol resulting in an increase in the cytosolic energy charge and [ATP]/[ADP] ratio. In contrast to these results, however, Stitt *et al.* (1982) have reported that although the cytosolic [ATP]/[ADP] ratio is considerably higher than that in the mitochondrial matrix or chloroplast stroma it actually decreases in the light; Stitt *et al.* (1982) suggest that previous measurements were due in part to contaminating mitochondrial and cytosolic material. This may be true since the actual level of this ratio in the cytosol will be dependent upon the rate of indirect ATP transfer by the phosphoglycerate/dihydroxy-acetone phosphate shuttle compared with ATP consumption in the cytosol. Cellular activities such as sucrose synthesis and protein synthesis are stimulated in the light and require ATP. Furthermore ATP transfer may well be limited either by its availability in the chloroplast or the capacity of the translocator. Such processes may indeed significantly reduce the estimated level of the [ATP]/[ADP] ratio.

Effects of light on mitochondrial metabolism

Effect of adenylate levels

As previously described, mitochondria are the sites of oxidation of the TCA cycle intermediates and ATP synthesis. Adenine nucleotides are intimately involved in these processes and control many of the respiratory enzymes (see Graham & Chapman, 1979). The oxidation of TCA intermediates by the respiratory chain appears to be controlled by the phosphorylation potential (Chance & Williams, 1956). Low phosphorylation potentials stimulate electron transport activity whilst high phosphorylation potentials inhibit (Moore, 1978). Mitochondria are, however, capable of synthesising ATP under a very high cytosolic phosphorylation potential (Moore & Bonner, 1981). The [ATP]/[ADP] translocator is important in this respect since the specificity for external ADP tends to keep the intramitochondrial phosphorylation potential low compared with the high cytosolic potential. A knowledge of the level of the cytosolic potential is therefore very important in determining the activity of the mitochondrial respiratory chain.

From the preceding section it is apparent that illumination results in an increase in the cytosolic [ATP]/[ADP] ratio (see Heber & Santarius, 1970; however, see also Stitt et al., 1982). Whether this ratio is sufficiently high to induce respiratory inhibition is uncertain. Obviously ATP consumption by cytosolic reactions will tend to decrease this ratio and favour respiratory ATP production. The finding that mitochondria can synthesise ATP against high cytosolic phosphorylation potentials (Moore & Bonner, 1981) would tend to suggest that even at high cytosolic [ATP]/[ADP] ratios, respiratory inhibition may be much less than previously anticipated (Heber, 1974, 1975). Indeed, the extent of inhibition of the respiratory chain in vivo in the light is unclear and reports vary from as much as 100% (Chevallier & Douce, 1976) to as little as 15% (Strotmann & Murakami, 1976). It is hoped that current work in our laboratory on direct measurements of mitochondrial respiratory activity in vivo using membrane potential probes may help to clarify some of these contradictory observations (Valles, Beechey & Moore, 1982). However, whatever the outcome of the studies there is an increasing amount of evidence in favour of some degree of respiratory activity in the light. Whether this activity is associated with ATP synthesis is uncertain.

In addition to the effects of adenine nucleotides on the respiratory chain itself, regulatory enzymes of the catabolic pathways such as phosphofructo-kinase, pyruvate dehydrogenase and isocitrate dehydrogenase may be subject to control by AMP, ADP or ATP (Turner & Turner, 1980). Similarly malate dehydrogenase and isocitrate dehydrogenase are inhibited by a high ratio of reduced to oxidised pyridine nucleotides. Some interaction with the dehydrogenase steps of the pentose phosphate pathway may also occur since most of these enzymes are subject to control by either adenylates or the adenine nucleotides. Studies with algae would tend to suggest that this pathway is operative in the light (Raven, 1976).

Effect on carbon metabolism

Many early biochemical studies seem to suggest that photosynthetically incorporated $^{14}CO_2$ does not readily appear to enter the TCA cycle, nor is pyruvate oxidation easily achieved (see Graham & Chapman, 1979). The finding that extinguishing the light results in a rapid labelling of intermediates would seem to suggest that TCA cycle and hence mitochondrial respiration is inhibited in the light. However, later experiments (Chapman & Graham, 1974; Graham, 1980), where ^{14}C-labelled organic acids were fed to mung bean leaves, indicated that the TCA cycle functions in the light at the same rate as it does in the dark except for a brief initial inhibition on transition from dark to light. In an investigation of glutamate metabolism in leaf discs (Jordan & Givan, 1979) it was found that although TCA cycle reactions continued operating in the light, the percentage of glutamate metabolised through them was reduced.

Earlier work on the effect of light on higher plant glycolysis showed that pyruvate oxidation was decreased upon illumination, possibly due to increased cytosolic $[ATP]/[ADP]$ ratios inhibiting enzymes such as phosphofructokinase, 3-phosphoglycerate kinase and pyruvate kinase (Turner and Turner, 1980). However, it has been suggested that such an inhibition only refers to illumination for a few minutes following darkness. This would agree with the suggestion (Graham & Chapman, 1979) that there is an initial inhibition of respiratory activity which is quickly followed by resumption of a rate similar to that in the dark. Chapman & Graham (1974) showed that rapid readjustments in metabolic pools occur during the first few minutes of illumination, and that these result in an apparent inhibition of the TCA cycle. Following establishment of new steady-state levels of intermediates, the cycle operates at a rate similar to that in the dark, and therefore the results from short-term studies tend to be misleading with respect to the extent of respiratory activity.

Relatively little work has been done on carbon and energy flow in C_4 plants although the pathway of C_4 photosynthesis requires rapid metabolite flow both between different compartments such as the mesophyll and bundle sheath cells and between different organelles (Coombs, 1976; Graham, 1980). Chapman & Osmond (1974) have presented some evidence for movement of carbon from mitochondria to chloroplasts. In particular, it was found that a large light-dependent transfer of label from intermediates of the TCA cycle to photosynthetic products was a characteristic of C_4 plants.

On the basis of this evidence Graham & Chapman (1979) have concluded that the TCA cycle continues to operate in the light.

Effect of photorespiration

In a discussion of the interactions between mitochondria and chloroplasts it is of importance to consider the interactions between photorespiratory carbon metabolism and mitochondrial respiration.

Photorespiration is a curious metabolic pathway in which carbon dioxide is released from previously fixed intermediates of the Calvin cycle that occurs in most crop plants (Chollet, 1977; Tolbert, 1979, 1980; Zelitch, 1979). It is a light-dependent process and is considered to be due to the formation in the chloroplast of the two-carbon compound glycolic acid and its subsequent oxidative decarboxylation in the leaf cytoplasm to yield carbon dioxide. Glycolate is oxidised to glyoxylate by the flavoprotein enzyme glycolate oxidase which is located in the peroxisomes. Glyoxylate is aminated to glycine by specific peroxisomal transaminases, and the glycine is transported into mitochondria. Glycine decarboxylation to serine, which occurs in the mitochondrial matrix, is currently considered to be the major source of photorespiratory carbon dioxide although other pathways may also operate. Serine can be further metabolised in the peroxisomes to glycerate which upon phosphorylation in the chloroplast can re-enter the Benson–Calvin cycle as phosphoglycerate. Thus the current scheme of the biochemistry of photorespiration involves extensive shuttling of metabolites between chloroplasts, peroxisomes and mitochondria, and it may be supposed that a close interaction must exist between the carbon metabolism of photorespiration and mitochondrial respiration.

Glycine decarboxylase is a multi-enzyme complex located in the mitochondrial matrix, and in addition to requiring tetrahydrofolate and pyridoxal phosphate as cofactors, it also requires NAD^+. The net products of this reaction from two glycines are one serine, one carbon dioxide, one ammonia and NADH. In isolated mitochondria NAD^+ reduced by this complex can be re-oxidised by the NADH dehydrogenase system of the respiratory chain (Moore et al., 1977). Thus the oxidative decarboxylation of glycine is linked to ATP synthesis *in vitro*. Obviously the continued glycine decarboxylation *in vivo* requires the regeneration of the NAD^+. It is not clear at present whether this is achieved by the respiratory chain or whether it is coupled to other enzymes such as glutamate dehydrogenase or NADH-hydroxypyruvate reductase (Tolbert, 1979). Tolbert (1979) has suggested that during glycine decarboxylation neither ammonia nor NADH should accumulate, as mitochondrial respiration is inhibited in the light. If mitochondrial respiratory activity is not inhibited in the light, it is conceivable that the NAD^+ is regenerated via electron transport activity, suggesting that ATP is synthesised *in vivo* during the decarboxylation reaction (Valles et al., 1982). Should respiratory activity be low then other mechanisms such as coupling the system to oxaloacetate reduction with eventual transfer of NADH to the cytosol via metabolite shuttles, must be sought (Woo & Osmond, 1976). The answer to this question has considerable implications for the study of photorespiration.

Mechanism of operation

The mechanism by which the TCA cycle operates in the light is perhaps worth considering at this point. From the preceding discussions it is apparent that upon illumination cytosolic [ATP]/[ADP] ratios increase due to the action of the phosphoglycerate/dihydroxyacetone phosphate shuttle, whilst there is relatively little change in the [NADH]/[NAD] levels although they are significantly lower than in the chloroplast stromal phase. Rates of cytosolic ATP consumption upon illumination have been reported to be approximately 40 μmol mg^{-1} chlorophyll h^{-1} (Giersch *et al.*, 1980*b*), suggesting that considerable secondary photosynthetic activity occurs in the cytosol such as sucrose synthesis. If this ATP consumption is sufficient to decrease the overall cytosolic phosphorylation potential little mitochondrial respiratory inhibition will be apparent. Respiratory activity would in turn allow TCA cycle operation. If, however, cytosolic ATP consumption is insufficient to prevent respiratory inhibition in the light then some other mechanism to allow TCA cycle operation must be sought. It is of course conceivable that respiratory activity is not an all or nothing process. Since mitochondrial respiration in isolated mitochondria appears to be inversely related to the phosphorylation potential (Moore, 1978) it is not difficult to construct a model in which respiration is somewhere between a fully inhibited and a stimulated state. Alternatively plant mitochondria do possess a branched non-phosphorylating electron transport chain which may allow the continued oxidation of TCA cycle intermediates under conditions of high cytosolic adenylate concentrations. The possession of a cyanide- and antimycin-resistant alternative oxidase has significance in this respect and several reports suggest that intermediates may be oxidised by a chain which completely by-passes all of the energy-conserving steps (Palmer, 1976, 1979; Rustin & Moreau, 1979; Moore & Rich, 1980). It may therefore be concluded that although mitochondrial respiration by the conventional cytochrome oxidase pathway may be controlled by the cytosolic phosphorylation potential, the presence of a non-phosphorylating electron transport system may allow turnover of the TCA cycle in order to supply carbon skeletons for biosynthetic purposes such as porphyrin and amino acid synthesis. The actual extent to which this occurs will of course be dependent on the activity of the non-phosphorylating pathways which may vary from tissue to tissue, but such ideas may help to explain the contradictory observations made by different groups of workers on the degree of TCA cycle activity in the light.

Concluding remarks

It is probably very apparent to the reader that interactions between photosynthesis and respiration are very complex and that we are still unsure as to the degree of mitochondrial respiration under illuminated conditions. The answer to this problem appears to rest in the levels of adenine and pyridine nucleotides both in the cytosol and the individual organelle compartments. These in their turn are not only dependent upon their utilisation in the cytosol for biosynthetic purposes but also on the relative rates of carbon export and import from the organelles in question. If ATP and pyridine nucleotide utilisation in the cytosol exceeds export from the chloroplast this will lower controlling factors such as the phosphorylation potential and result in mitochondrial respiratory activity, with the proviso that it is in equilibrium with the cytosolic phosphorylation potential. Obviously a lot of 'ifs' still have to be resolved. The body of evidence in favour of continued TCA cycle activity in the light is increasing and should mitochondrial respiratory chain activity via cytochrome oxidase be insufficient to account for this turnover other mechanisms must be sought. Perhaps the presence of non-phosphorylating pathways may provide an answer to this question. Whatever the outcome it is obvious that further experimental knowledge of intracellular carbon and energy transport is required for a fuller understanding of the complex interaction between mitochondria and chloroplasts in the integrated cell.

The author gratefully acknowledges Professors W. D. Bonner Jr and R. B. Beechey for their encouragement and the Royal Society, Science Research Council and Agricultural Research Council for financial support.

References

Arnon, D. I., Allen, M. B. & Whatley, F. R. (1954). Photosynthesis by isolated chloroplasts. *Nature, London*, **174**, 394–6.

Atkinson, D. E. (1977). *Cellular Energy Metabolism and its Regulation*. New York & London: Academic Press.

Chance, B. & Williams, G. R. (1956). The electron transport chain and oxidative phosporylation. *Advances in Enzymology*, **17**, 65–98.

Chapman, E. A. & Graham, D. (1974). The effect of light on the tricarboxylic acid cycle in green leaves. II. Intermediary metabolism and the location of control points. *Plant Physiology*, **53**, 886–92.

Chapman, E. A. & Osmond, C. B. (1974). The effect of light on the tricarboxylic acid cycle in green leaves. III. A comparison between some C_3 and C_4 plants. *Plant Physiology*, **53**, 893–8.

Chevallier, D. & Douce, R. (1976). Interactions between mitochondria and chloroplasts in cells. I. Action of cyanide and of 3-(3,4-dichlorophenyl)-

1,1-dimethylurea on the spore of *Funaria hygrometrica*. *Plant Physiology*, **57**, 400–2.

Chollet, R. (1977). Biochemistry of photorespiration. *Trends in Biochemical Sciences*, **2**, 155–9.

Coombs, J. (1976). Interactions between chloroplasts and cytoplasm in C_4 plants. In *The Intact Chloroplast*, ed. J. Barber, pp. 279–313. Amsterdam: Elsevier.

Day, D. A. & Wiskich, J. T. (1974). The oxidation of malate and exogenous NADH by isolated plant mitochondria. *Plant Physiology*, **53**, 104–9.

Douce, R., Mannella, C. A. & Bonner, W. D. (1973). The external NADH dehydrogenases of intact plant mitochondria. *Biochimica et Biophysica Acta*, **292**, 105–16.

Giersch, C. H., Heber, U. & Krause, G. H. (1980*a*). ATP transfer from chloroplasts to the cytosol of leaf cells during photosynthesis and its effect on leaf metabolism. In *Plant Membrane Transport: Current Conceptual Issues*, ed. R. M. Spanswick, W. J. Lucas & J. Dainty, pp. 65–79. Amsterdam: Elsevier.

Giersch, C., Heber, U., Kobayashi, Y., Inoue, Y., Shibata, K. & Heldt, H. W. (1980*b*). Energy charge, phosphorylation potential and proton motive force in chloroplasts. *Biochimica et Biophysica Acta*, **590**, 59–73.

Graham, D. (1980). Effects of light on dark respiration. In *Biochemistry of Plants*, vol. 2, ed. D. D. Davies, pp. 526–80. New York & London: Academic Press.

Graham, D. & Chapman, E. A. (1979). Interactions between photosynthesis and respiration in higher plants. In *Encyclopedia of Plant Physiology, Photosynthesis II*, vol. 6, ed. M. Gibbs & E. Latzko, pp. 150–62. Berlin: Springer-Verlag.

Halliwell, B. (1978). The chloroplast at work. *Progress in Biophysics and Molecular Biology*, **33**, 1–54.

Halliwell, B.(1981). *Chloroplast Metabolism*. Oxford: Clarendon Press.

Hampp, R., Goller, M. & Ziegler, H. (1982). Adenylate levels, energy charge and phosphorylation potential during dark–light and light–dark transition in chloroplasts, mitochondria and cytosol of mesophyll protoplasts from *Avena sativa* L. *Plant Physiology*, **69**, 448–55.

Heber, U. (1974). Metabolite exchange between chloroplasts and cytoplasm. *Annual Review of Plant Physiology*, **25**, 393–421.

Heber, U. (1975). Energy transfer within leaf cells. In *Proceedings of the Third International Congress on Photosynthesis*, vol. 2, ed. M. Avron, pp. 1335–48. Amsterdam: Elsevier.

Heber, U., Hallier, U. W. & Hudson, M. A. (1967). Lokalisation von Enzymen des reduktiven und oxydativen Pentosephosphate-Zyklus in den Chloroplasten und Permeabilität der Chloroplasten-Membran gegenuber Metaboliten. *Zeitschrift für Naturforschung*, **22b**, 1200–15.

Heber, U. & Heldt, H. W. (1981). The chloroplast envelope. *Annual Review of Plant Physiology*, **32**, 139–68.

Heber, U. & Krause, G. H. (1971). Transfer of carbon, phosphate energy and reducing equivalents across the chloroplast envelope. In *Photosynthesis and Photorespiration*, ed. M. D. Hatch, C. B. Osmond & R. O. Slatyer, pp. 218–25. New York: Wiley-Interscience.

Heber, U. & Krause, G. H. (1972). Hydrogen and proton transfer across the chloroplast envelope. In *Progress in Photosynthesis*, vol. II, ed. G. Forti, M. Avron & A. Melandri, pp. 1023–33. The Hague: Junk N.V.

Heber, U. & Santarius, K. A. (1970). Direct and indirect transfer to ATP and ADP across the chloroplast envelope. *Zeitschrift für Naturforschung*, **25b**, 718–28.

Heber, U. & Walker, D. A. (1979). The chloroplast envelope – barrier or bridge? *Trends in Biochemical Sciences*, **4**, 252–6.

Heldt, H. W. (1969). Adenine nucleotide translocation in Spinach chloroplasts. *FEBS Letters*, **5**, 11–14.

Heldt, H. W. (1976a). Metabolite carriers of chloroplasts. In *Encyclopedia of Plant Physiology, Transport in Plants III*, vol. 3, ed. C. R. Stocking & U. Heber, pp. 137–43. Amsterdam: Elsevier.

Heldt, H. W. (1976b). Metabolite transport in spinach chloroplasts. In *The Intact Chloroplast*, ed. J. Barber, pp. 215–34. Amsterdam: Elsevier.

Heldt, H. W. & Sauer, F. (1970). The inner membrane of the chloroplast envelope as the site of specific metabolite transport. *Biochimica et Biophysica Acta*, **234**, 83–91.

Heldt, H. W., Werdan, K., Milovancev, M. & Geller, G. (1973). Alkalisation of the chloroplast stroma caused by light dependent proton flux into the thylakoid space. *Biochimica et Biophysica Acta*, **314**, 224–41.

Jordan, B. R. & Givan, C. V. (1979). Effects of light and inhibitors on glutamate metabolism in leaf discs of *Vicia faba* L. *Plant Physiology*, **64**, 1043–7.

Klingenberg, M. (1970a). Mitochondrial metabolite transport. *FEBS Letters*, **6**, 145–54.

Klingenberg, M. (1970b). Metabolite transport in mitochondria: an example for intracellular membrane function. *Essays in Biochemistry*, **6**, 5–159.

Klingenberg, M. (1979). The ADP,ATP shuttle of the mitochondrion. *Trends in Biochemical Sciences*, **4**, 249–52.

Kobayashi, Y., Inoue, Y., Furuya, F., Shibata, K. & Heber, U. (1979). Regulation of adenylate levels in intact spinach chloroplasts. *Planta, Berlin*, **147**, 69–75.

Krause, G. H. & Heber, U. (1976). Energetics of intact chloroplasts. In *The Intact Chloroplast*, ed. J. Barber, pp. 171–214. Amsterdam: Elsevier.

La Noue, K. F., Walajtys, E. I. & Williamson, J. R. (1973). Regulation of glutamate metabolism and interaction with the citric acid cycle in rat heart mitochondria. *Journal of Biological Chemistry*, **248**, 7171–83.

Lilley, R. McC., Chon, C. J., Mosbach, A. & Heldt, H. W. (1977). The distribution of metabolites between spinach chloroplasts and medium during photosynthesis *in vitro*. *Biochimica et Biophysica Acta*, **460**, 259–72.

Mazliak, P. (1973). Lipid metabolism in plants. *Annual Review of Plant Physiology*, **24**, 287–310.

Moore, A. L. (1978). The electrochemical gradient of protons as an intermediate between electron transport and ATP synthesis. In *Plant Mitochondria*, ed. G. Ducet & C. Lance, pp. 85–92. Amsterdam: Elsevier/North-Holland.

Moore, A. L. & Bonner, W. D. (1981). A comparison of the phosphorylation potential and electrochemical proton gradient in mung bean mitochondria and phosphorylating submitochondrial particles. *Biochimica et Biophysica Acta*, **634**, 117–28.

Moore, A. L., Jackson, C., Halliwell, B., Dench, J. E. & Hall, D. O. (1977). Intramitochondrial localisation of glycine decarboxylase in spinach leaves. *Biochemical and Biophysical Research Communications*, **78**, 483–91.

Moore, A. L. & Rich, P. R. (1980). The bioenergetics of plant mitochondria. *Trends in Biochemical Sciences*, **5**, 284–8.

Palmer, J. M. (1976). The organisation and regulation of electron transport in plant mitochondria. *Annual Review of Plant Physiology*, **27**, 133–57.

Palmer, J. M. (1979). The uniqueness of plant mitochondria. *Biochemical Society Transactions*, **7**, 246–52.

Palmieri, F., Stipani, I., Quagliariello, E. & Klingenberg, M. (1972). Kinetic study of the tricarboxylate carrier in rat liver mitochondria. *European Journal of Biochemistry*, **26**, 587–94.

Pfaff, E., Heldt, H. W. & Klingenberg, M. (1969). Kinetics of the adenine nucleotide exchange. *European Journal of Biochemistry*, **10**, 484–93.

Raven, J. A. (1976). Division of labour between chloroplast and cytoplasm. In *The Intact Chloroplast*, ed. J. Barber, pp. 403–43. Amsterdam: Elsevier.

Rustin, P. & Moreau, F. (1979). Malic enzyme activity and cyanide insensitive electron transport in plant mitochondria. *Biochemical and Biophysical Research Communications*, **88**, 1125–31.

Santarius, K. A. & Heber, U. (1965). Changes in the intracellular levels of ATP, ADP, AMP and P_i and regulatory function of the adenylate system in leaf cells during photosynthesis. *Biochimica et Biophysica Acta*, **102**, 39–54.

Schnarrenberger, C. & Fock, H. (1976). Interactions among organelles involved in photorespiration. In *Encyclopedia of Plant Physiology, Transport in Plants III*, vol. 3, ed. C. R. Stocking & U. Heber, pp. 185–234. Berlin: Springer-Verlag.

Stitt, M., Lilley, R. McC. & Heldt, H. W. (1982). Adenine nucleotide levels in the cytosol, chloroplasts and mitochondria of wheat leaf protoplasts. *Plant Physiology*, **70**, 965–70.

Stitt, M., Wirtz, W. & Heldt, H. W. (1980). Metabolite levels during induction in the chloroplast and extrachloroplast compartments of spinach protoplasts. *Biochimica et Biophysica Acta*, **592**, 85–102.

Strotmann, H. & Murakami, S. (1976). Energy transfer between cell compartments. In *Encyclopedia of Plant Physiology, Transport in Plants III*, vol. 3, ed. C. R. Stocking & U. Heber, pp. 398–418. Berlin: Springer-Verlag.

Tolbert, N. E. (1979). Glycolate metabolism by higher plants and algae. In *Encyclopedia of Plant Physiology, Photosynthesis II*, vol. 6, ed. M. Gibbs & E. Latzko, pp. 338–52. Berlin: Springer-Verlag.

Tolbert, N. E. (1980). Photorespiration. In *Biochemistry of Plants*, vol. 2, ed. D. D. Davies, pp. 488–525. New York & London: Academic Press.

Turner, J. F. & Turner, D. H. (1980). The regulation of glycolysis and pentose phosphate pathway. In *Biochemistry of Plants*, vol. 2, ed. D. D. Davies, pp. 279–316. New York & London: Academic Press.

Valles, K. L. M., Beechey, R. B. & Moore, A. L. (1982). Membrane potential measurements in intact leaf protoplasts. In *Second European Bioenergetics Conference*, pp. 329–30.

Vignais, P. V., Douce, R., Lauquin, G. J. M. & Vignais, P. M. (1976). Binding of radioactively labelled carboxyactractyloside, actractyloside and bongkrekic acid to the ADP translocator of potato mitochondria. *Biochimica et Biophysica Acta*, **440**, 688–96.

Wiskich, J. T. (1977). Mitochondrial metabolite transport. *Annual Review of Plant Physiology*, **28**, 45–69.

Wiskich, J. T. (1980). Control of the Krebs cycle. In *Biochemistry of Plants*, vol. 2, ed. D. D. Davies, pp. 244–78. New York & London: Academic Press.

Woo, K. C. & Osmond, C. B. (1976). Glycine decarboxylation in mitochondria isolated from spinach leaves. *Australian Journal of Plant Physiology*, **3**, 771–85.

Zelitch, I. (1979). Photorespiration: studies with whole tissue. In *Encyclopedia of Plant Physiology, Photosynthesis II*, vol. 6, ed. M. Gibbs & E. Latzko, pp. 353–67. Berlin: Springer-Verlag.

D. D. DAVIES

13 The co-ordination and integration of metabolic pathways

The biochemical approach to the study of metabolism involves the grinding, crushing or smashing of cells to reduce a metabolic sequence to its components and finally an attempt to determine how the components interact to produce a system. The dangers inherent in coupling a physically destructive reduction to a philosophically dubious reconstruction have been illustrated by parables concerning motor-cars. D. E. Green suggested that a Martian mechanic might assemble the components of a disintegrated car to produce a vacuum cleaner and deduce that the function of the car was to clean carpets! Krebs & Veech (1969) suggested that the demented giant of a biochemist who homogenised a car in an aqueous environment and noted an evolution of gas might conclude that the car is driven by the gas produced by the action of the battery acid on metal.

It is perhaps significant that the tellers of these parables contributed extensively to this approach and their achievement is enshrined in the metabolic maps so elaborately produced as posters by the drug houses. The complexity of these maps constitutes a testimony to the dedication of metabolic biochemists, but they present a terrifying picture to students, creating despair in terms of memory and comprehension. The map of Britain gives little information about the state of the economy and a metabolic map gives little idea of the well-being of an organism. To understand the economics of the cell we need quantitative information about the fluxes through metabolic pathways and the principles of integration and control which produce a functional organism. We must recognise that 'A process cannot be understood by stopping it. Understanding must move with the flow of the process, must join it and flow with it' (Herbert, 1965).

Early consideration of the dynamic properties of biological systems was concerned with rate processes. Liebig (1855) appears to have been the first to formulate a law concerning limiting factors: 'Crop yield is proportional to the components in the soil which are present in smallest quantity' (cited in Romell, 1924). Blackman (1905) proposed that the rate of a complex process is 'limited by the pace of the slowest factor'. Krebs & Kornberg (1957) recognised the incompatibility between a slowest reaction and a steady state in which all reactions proceed at the same pace; they therefore suggested that one reaction – the *pacemaker* – might determine that pace.

About this time, concepts such as metabolic control by feedback and the significance of allosteric enzymes were evolving (see Wolstenholme & O'Connor, 1959), so that a molecular basis for metabolic control was quickly established. For example, phosphofructokinase, which plays a major role in controlling the flux through the glycolytic pathway, is an allosteric enzyme showing very complex kinetics and interactions with many metabolites. Thus the skeletal enzyme is inhibited by high concentrations of one of its substrates, ATP, and activated by one of its products, fructose bisphosphate, producing a positive feedback system which may explain the large and rapid increase in fructose bisphosphate which occurs when glycolysis is stimulated:

$$\text{Glucose-6-P} \rightleftharpoons \text{fructose-6-P} \xrightarrow[\text{ATP} \quad \text{ADP}]{\quad (+) \quad} \text{fructose-1,6-P}$$

However, this example contains a feature which argues against the concept of a pacemaker and supports the view that metabolic control is a property

Fig. 1. Effect of aldolase on the activity of phosphofructokinase. Activity of phosphofructokinase is measured by the change in E_{340} due to oxidation of NADH in the coupled reactions:

Fructose-6-phosphate + ATP
  → fructose bisphosphate + ADP
ADP + phosphoenolpyruvate ⎫
  → ATP + pyruvate   ⎬ Coupled enzymes added for assay
Pyruvate + NADH + H$^+$ ⎭
  → NAD$^+$ + lactate

(After El-Badry, Otani & Mansour, 1973.)

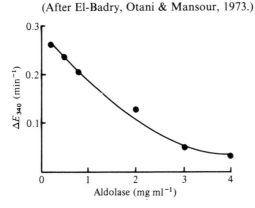

of the system rather than residing in a particular component. Consider a situation where the next enzyme in the glycolytic sequence – aldolase – is highly active. Fructose bisphosphate would be removed as soon as it was formed, and the positive feedback loop eliminated; hence, high concentrations of aldolase would reduce the activity of phosphofructokinase (Fig. 1) and for effective metabolic control the activities of phosphofructokinase and aldolase have to be matched. There do not appear to have been comparable experiments done with plants, but the correlation between the respiratory activity of carrot discs and their concentration of fructose bisphosphate (+ triose phosphate) is at least consistent with the proposal (Fig. 2).

The general proposition that metabolic control is a system's property is now widely accepted (Kacser & Burns, 1973, 1979) and systems analysis has been successfully applied in studies of the glycolytic flux. With the steady availability of minicomputers, we can anticipate rapid developments in the analysis of the major pathways of metabolism. It is convenient to treat the pathways as separate entities so that the control of one sequence is considered to be independent of metabolic events in another sequence. Current research, however, is concerned with the interaction between pathways and two aspects will be considered here.

Metabolic interlock

A metabolite in one pathway may be an effector in another pathway and also a substrate in another pathway, so that the fluxes through the pathways are interlocked (Kane *et al.*, 1971). For example, phosphoenol-pyruvate is a substrate for at least six enzymes (Fig. 3), and thus interlocks the

Fig. 2. Correlation between rate of respiration and concentration of fructose bisphosphate (FDP) and triose phosphate (TP) in carrot tissue. (After Everson & Rowan, 1965.)

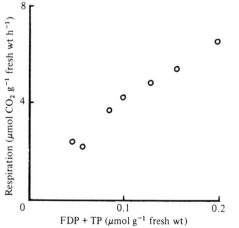

metabolic pathways of glycolysis, gluconeogenesis, the tricarboxylic acid cycle and aromatic biosynthesis. Because phosphoenolpyruvate is a co-substrate with erythrose-4-phosphate for entry into the aromatic pathway, there is also an additional interlock with the pentose phosphate pathway. These interlocks at the substrate level are reinforced by interactions involving phosphoenolpyruvate as an allosteric effector (Fig. 4).

Metabolic buffering

In a steady state the flux through the system is constant and the concentrations of all intermediates are constant. Studies on metabolic control have tended to concentrate on the rapid changes of flux in the glycolytic pathway in response to anoxia, but the physiological maintenance of a steady state, i.e. homeostasis, is equally important. Failure to recognise this can produce misunderstandings. For example, Masters (1977), considering the changes in metabolite levels which occur following anoxia, noted 'only a small percentage change in the levels of ATP after 25 seconds of anoxia – hardly the rapid and dramatic effect which one might expect as consistent with the wide generalizations which have been sponsored in favour of the role of energy charge in metabolic control'. It is a little like the suggestion that because a constant-temperature bath was not showing wild fluctuations in temperature, the thermostat could not have been responding to changes in temperature!

If we consider that homeostasis is beneficial to the organism, then

Fig. 3. Reactions involving phosphoenolpyruvate (PEP) in plants. EC 2.7.1.40, pyruvate kinase; 2.7.9.1, pyruvate phosphate dikinase; 4.1.1.31, PEP carboxylase; 4.1.1.49, PEP carboxykinase; 4.1.2.15, phospho-2-keto-3-deoxy-heptonate aldolase; 4.2.1.11, enolase; ?, pyruvylshikimate phosphate synthetase. (After Davies, 1979.)

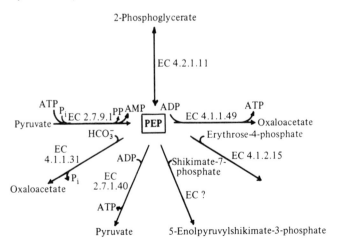

components which are involved in many reactions will need to be particularly carefully controlled, and I will identify the following as of special importance.

$$\frac{[ATP]}{[ADP][P_i]}, \quad \frac{[NAD]}{[NADH]}, \quad \frac{[NADP]}{[NADPH]} \quad \text{and pH.}$$

If these metabolites are maintained at constant concentration, then the reactions involving them must be fast, producing near-equilibrium conditions. Holzer, Schultz & Lynen (1956) pointed out that if the pyridine nucleotide dehydrogenases are at near-equilibrium then

$$\frac{[\text{Oxidised substrate}] \times [\text{NAD(P)H}_2]}{[\text{Reduced substrate}] \times [\text{NAD(P)}]} = K$$

or

$$\frac{[\text{NAD(P)}]}{[\text{NAD(P)H}_2]} = \frac{1}{K_{eq}} \times \frac{[\text{oxidised substrate}]}{[\text{reduced substrate}]}.$$

Fig. 4. Metabolic interactions involving phosphoenolpyruvate (PEP). Metabolic pathways are indicated by solid bars, control interactions by dashed lines. (+) is activation; (−) is inhibition. F-6-P, fructose-6-phosphate; G-1-P and G-6-P, glucose-1-phosphate and glucose-6-phosphate.

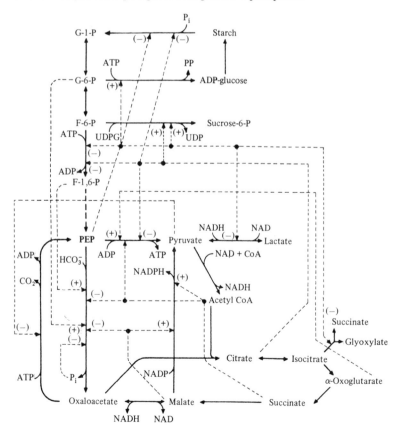

The ratio [NAD]/[NADH] in the cytoplasm can be measured from the [lactate]/[pyruvate] ratio, the [NAD]/[NADH] ratio in the mitochondria from the [glutamate]/[α-oxoglutarate] $\times$ [NH_4^+] ratio and the ratio [NADP]/[NADPH] in the cytoplasm from the [malate]/[pyruvate] $\times$ [CO_2] ratio. The presence of an acid-storing vacuole prevents this equilibrium method being applied to plants and a direct assay of the nicotinamide adenine dinucleotides is necessary (Table 1). Assuming that the link between the cytoplasmic NAD couple and the phosphorylation state of the adenine nucleotide ratio ([ATP]/[ADP] $\times$ [P_i]) is near equilibrium due to the activities of triose phosphate dehydrogenase and 3-phosphoglycerate kinase, then the phosphorylation state of the adenine nucleotides can be measured from the relationship

$$\frac{[ATP]}{[ADP] \times [P_i]} = \frac{[pyruvate]}{[lactate]} \times \frac{[glyceraldehyde\ phosphate]}{[3\text{-phosphoglycerate}]} \times K_{eq}.$$

The calculated adenine nucleotide ratio for the cytosol of rat liver is c. 550, which agrees well with the measured ratio for whole liver.

Just as the adenine nucleotide ratio appears to be linked to the NAD couple so the NAD couple is linked to the NADP couple in the cytosol because lactate dehydrogenase and malic enzyme share a common reactant. Thus

$$\begin{aligned}
\text{Pyruvate} &= K_{l.d.} \times \frac{[lactate] \times [NAD]}{[NADH_2]} \\
&= K_{m.e.} \times \frac{[malate] \times [NADP]}{[NADPH_2] \times [CO_2]},
\end{aligned}$$

where l.d. and m.e. represent lactate dehydrogenase and malic enzyme respectively.

Thus all three ratios are interlocked with one another and with the substrates of the various enzymes involved, producing a heavily metabolite-buffered system. It should be noted that in these equilibrium reactions the nucleotides are present at much lower concentrations than the other reactants. Hence, a small change in the nucleotide ratio will produce an equal change

Table 1. *Pyridine nucleotide ratios*

Tissue	Method of assay	$\frac{NAD}{NADH}$	$\frac{NADP}{NADPH}$	Reference
Rat liver	Freeze-clamp			Krebs (1973)
	(*a*) Cytoplasmic ratio	1160	0.01	
	(*b*) Mitochondrial ratio	7.3	—	
Apple (pulp)	Fluorometric	10	1.0	Rhodes &
Pear (pulp)	Fluorometric	8	0.7	Wooltorton
				(1968)

in the ratio of the other substrate pair, but a large change in the *concentration* of the metabolites. It is for this reason that we consider the nucleotides as the primary regulators of metabolism, and a well-buffered system will show very small changes in the various nucleotide ratios. Since the various nucleotide pairs are linked together, it is difficult to see why a particular ratio such as energy charge should have any greater significance than the ratio [NAD]/[NADH], for if one changes then the others must also change! Before leaving this point, it should be noted that a basic assumption is that dehydrogenases in the cytosol are specific for their nicotinamide adenine dinucleotides. If this were not so, then the ratio [NAD]/[NADH] would have to be almost exactly equal to the ratio [NADP]/[NADPH]. Rhodes (1973) has suggested that in some plants (e.g. melon) alcohol dehydrogenase is active with [NAD]/[NADH] and with [NADP]/[NADPH]. If two forms of the enzyme are separated in different organelles there is no problem, but if a single enzyme with dual specificity is present in the cytosol it will tend to make [NAD]/[NADH] equal to [NADP]/[NADPH].

The kinetics of equilibrium reactions

Enzymologists have developed rate equations for the special conditions that at time t_0 the products are zero. This limitation makes the algebra easier but is a condition of little interest to a biochemical physiologist. *In vivo*, substrates and products will always be present and equilibrium kinetics may apply. A theoretical treatment has been provided by Anderson (1974).

Starting from the rate equation for an ordered bi bi enzyme

$$v = \frac{V_1 V_2 (AB - PQ/K_{eq})}{\begin{array}{c} K_1 + K_2 A + K_3 B + K_4 AB + K_5 P + K_6 Q + K_7 PQ \\ + K_8 AP + K_9 BQ + K_{10} ABP + K_{11} BPQ \end{array}}, \tag{1}$$

where V_1 and V_2 are maximal velocities in the forward and reverse directions; A and B are substrate concentrations; P and Q are product concentrations; K_{eq} is the equilibrium constant; and K_1 to K_{11} are collections of rate constants.

For the equilibrium condition, $PQ/AB = K_{eq}$ and $v = 0$; hence $AB = PQ/K_{eq}$. For a small perturbation y of one of the reactants, say A,

$$v = \frac{y V_1 V_2 AB}{K_{12} + K_{13} A + K_{13} y A}, \tag{2}$$

where K_{12} is a collection of A-independent constants and K_{13} is a collection of A-dependent constants.

For fixed values of A, B, P and Q but with varying y

$$v = \frac{K_{14} y}{K_{15} + y}, \tag{3}$$

where $K_{14} = V_1 V_2 B / K_{13}$ and $K_{15} = 1 + K_{12}/(K_{13} A)$.

Thus the relationship between v and the fractional perturbation y resembles the Michaelis–Menten relationship (Fig. 5) and leads to the important conclusion that for small perturbations the rate at which the reaction readjusts to equilibrium is independent of which reactant is changed.

The physiological significance of these considerations can be illustrated by the control of the cytoplasmic pH. It has been suggested (Davies, 1973) that the pH of the cytoplasm is controlled by a metabolic pH-stat.

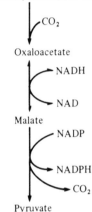

Glucose → Phosphoenolpyruvate

CO_2

Oxaloacetate

NADH

NAD

Malate

NADP

NADPH

CO_2

Pyruvate

The concentration of malic acid in the cytoplasm depends on the balance between carboxylation and decarboxylation (Fig. 6). If the pH of the

Fig. 5. Velocity versus fractional perturbation of A drawn for $K_{12} = 2 + K_{13}A$. The figure illustrates eqn. (3) for an arbitrarily assumed set of initial conditions in which the A-dependent terms account for one-third the total magnitude of the denominator of eqn. (1). This assumption affects only the location of the numerical values along the abscissa; for other initial conditions the position of K_{15} on the abscissa is always to the right of $y = 1.0$. (After Anderson, 1974.)

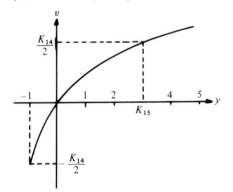

cytoplasm goes alkaline, phosphoenolpyruvate carboxylase will increase in activity leading to an increased production of malic acid, whilst malic enzyme will decrease in activity reducing the rate of malate removal so that the malic acid concentration would be expected to titrate the cytoplasm back to the controlled value (presumed to be close to pH 7). This model assumes that the malic enzyme catalyses a non-equilibrium reaction, but the sigmoid kinetics of malic enzyme are at variance with this proposal (Fig. 7). An increased cytoplasmic pH will decrease the activity of malic enzyme, but the increased

Fig. 6. A metabolic pH-stat based on control by malic enzyme and phosphoenolpyruvate (PEP) carboxylase. (After Davies, 1973.)

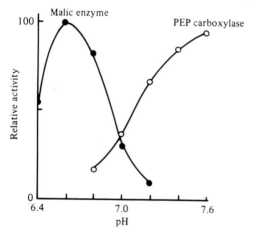

Fig. 7. Effect of pH and malate concentration on the rate of malate decarboxylation catalysed by partially purified 'malic' enzyme. (*a*) Direct plot of rate versus malate concentration; (*b*) double-reciprocal plot; (*c*) Hill plot. ○, pH 6.9 (nH = 1); △, pH 7.3 (nH = 2.0); □, pH 7.6 (nH = 2.1). (After Davies & Patil, 1974.)

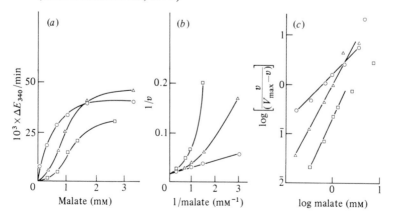

concentration of malic acid produced by carboxylation will greatly increase the rate of malate removal because the rate equation for malate removal is given by

$$v = \frac{V_{max} \times [malate]^2}{K_m + [malate]^2},$$

When the malate concentration is low compared with K_m, a doubling of the malate concentration will lead to a 4-fold increase in the rate of malate removal, so that the possession of sigmoid kinetics by malic enzyme leads to a self-defeating system in terms of the proposed metabolic pH-stat.

If, however, we assume that the malic enzyme catalyses an equilibrium reaction, the anomaly disappears. Thus an increase in pH will cause an increased production of malate via phosphoenolpyruvate carboxylase but very little removal of malate by malic enzyme because the equation

$$K_{eq} = \frac{[malate] \times [NADP]}{[pyruvate] \times [CO_2] \times [NADPH]}$$

predicts that if the malate concentration is large relative to NADP, equilibrium will tend to be restored by changes in the [NADP]/[NADPH] ratio. However, the ratio [NADP]/[NADPH] is coupled to the [NAD]/[NADH] and [ATP]/[ADP] ratios so that there will be a buffering action tending to block the removal of malate. Furthermore, another consideration leads to the prediction that under alkaline conditions, malic enzyme in the cytoplasm will synthesise malic acid. This is because the reaction catalysed by malic enzyme does not involve the production of a proton. However, most other dehydrogenases involve the production of a proton

$$XH_2 + NADP^+ \rightleftharpoons X + NADPH + H^+,$$

so that alkaline conditions will tend to increase the ratio [NADPH]/[NADP] and malic enzyme will respond to this ratio by producing more malate. Again it should be noted that a small change in the ratio [NADPH]/[NADP] will lead to a large change in the *amount* of malic acid.

The majority view about the role of malic enzyme in plants appears to be that it catalyses a non-equilibrium reaction. However, some data consistent with the proposition that it catalyses an equilibrium reaction have been obtained by Miss Comley working in my laboratory. She supplied potato discs with $[1-^{14}C]$pyruvate or $[1-^{14}C]$glutamate and measured the incorporation of ^{14}C into malate under conditions where there was a net synthesis of malate. $[1-^{14}C]$pyruvate could lead to incorporation into malate by two routes:

(*a*) $[1-^{14}C]$pyruvate $+ CO_2 + NADPH \rightleftharpoons [1-^{14}C]$malate $+ NADP$

(*b*) $[1-^{14}C]$pyruvate $\rightarrow$ acetyl CoA $+ ^{14}CO_2$

$^{14}CO_2 + $ phosphoenolpyruvate $\rightarrow [4-^{14}C]$oxaloacetate $+ P_i$

$[4-^{14}C]$oxaloacetate $+ NADH + H^+ \rightarrow [4-^{14}C]$malate $+ NAD^+$.

On the other hand, ^{14}C from [1-^{14}C]glutamate can only enter malate after decarboxylation followed by the fixation of $^{14}CO_2$ via route (*a*) or (*b*). Hence we can compare the incorporation of ^{14}C into malate from [1-^{14}C]pyruvate and [1-^{14}C]glutamate and estimate the minimum fraction of malic acid formed via the malic enzyme (Table 2). This result is at least consistent with the equilibrium proposals of this paper, and implies that malic enzyme catalyses an equilibrium reaction.

The metabolic model outlined in this paper allows homeostasis to be maintained against a range of environmental changes, but no buffering system can be expected to cope with extreme changes. It is reasonable to suppose that plants have developed special strategies to cope with particular stresses. For example, many stress conditions produce enhanced protein degradation (Cooke, Oliver & Davies, 1979) and in many plants an osmotic stress leads to the accumulation of proline. This can be rationalised as follows. To withstand wilting a plant must produce osmotically active molecules in the cytoplasm, and to prevent leakage of these molecules into the vacuole the molecules would need to be charged. If these molecules were acids or bases, they would change the pH of the cytoplasm; so the osmotically active molecules need to be zwitterions, e.g. amino acids. However, most amino acids are relatively insoluble and therefore could not function to retain water in the cytoplasm. Proline, however, is a very soluble amino acid and this may be the reason why it is so widely produced under wilting conditions. Similarly, we can anticipate other adaptations to meet different stresses, but it will not be easy to predict the strategy which the plant has adopted to cope with a particular stress. Having read the book of Joshua, it might appear sensible to capture Jericho by sounding trumpets, but Joshua must have been either peculiarly naive or highly imaginative to believe that such a strategy would bring about the fall of Jericho!

Table 2. *Reversibility of malic enzyme in potato discs*

^{14}C-labelled compounds	^{14}C in malate (c.p.m.)	^{14}C in CO_2 (c.p.m.)
[1-^{14}C]glutamate	$x = 3865$	$y = 212090$
[1-^{14}C]pyruvate	$a = 16830$	$b = 785330$

Washed potato discs were supplied with [1-^{14}C]pyruvate and [1-^{14}C]glutamate and the incorporation of ^{14}C into malate and carbon dioxide determined.
Counts in malate due to malic enzyme $= a - bx/y = 2500$. Percentage counts in malate due to malic enzyme $= 15\%$.

References

Anderson, J. H. (1974). Control of enzymatic velocity under near-equilibrium conditions. *Journal of Theoretical Biology*, **47**, 153–61.

Blackman, F. F. (1905). Optima and limiting factors. *Annals of Botany*, **19**, 281–95.

Cooke, R. J., Oliver, J. & Davies, D. D. (1979). Stress and protein turnover in *Lemna minor*. *Plant Physiology*, **64**, 1109–13.

Davies, D. D. (1973). Control of and by pH. *Symposia of the Society for Experimental Biology*, **27**, 513–29.

Davies, D. D. (1979). The central role of phosphoenolpyruvate in plant metabolism. *Annual Review of Plant Physiology*, **30**, 131–58.

Davies, D. D. & Patil, K. D. (1974). Regulation of 'malic' enzyme of *Solanum tuberosum* by metabolites. *Biochemical Journal*, **137**, 45–53.

El-Badry, A. M., Otani, A. & Mansour, T. E. (1973). Studies on heart phosphofructokinase. Role of fructose-1,6-diphosphate in enzyme activity. *Journal of Biological Chemistry*, **248**, 557–63.

Everson, R. G. & Rowan, K. S. (1965). Phosphate metabolism and induced respiration in washed carrot slices. *Plant Physiology*, **40**, 1247–50.

Herbert, F. (1965). *Dune*. (Berkeley Medallion Edition, 1975.)

Holzer, H., Schultz, G. & Lynen, F. (1956). Bestimmung des Quotienten DPNH/DPN in lebenden Hefezellen durch Analyse stationärer Alkohol- und Acetaldehyd-Konzentrationen. *Biochemische Zeitschrift*, **328**, 252–63.

Kacser, H. & Burns, J. A. (1973). The control of flux. *Symposia of the Society for Experimental Biology*, **27**, 65–104.

Kacser, H. & Burns, J. A. (1979). Molecular democracy: who shares the controls? *Biochemical Society Transactions*, **7**, 1149–60.

Kane, J. F., Stenmark, S. L., Calhoun, D. H. & Jensen, R. A. (1971). Metabolic interlock. The role of the subordinate type of enzyme in the regulation of a complex pathway. *Journal of Biological Chemistry*, **246**, 4308–16.

Krebs, H. A. (1973). Pyridine nucleotides and rate control. *Symposia of the Society for Experimental Biology*, **27**, 299–318.

Krebs, H. A. & Kornberg, H. L. (1957). *Ergebnisse der Physiologie*, **49**, 212.

Krebs, H. A. & Veech, R. L. (1969). Equilibrium relations between pyridine nucleotides and their roles in the regulation of metabolic processes. *Advances in Enzyme Regulation*, **7**, 397–413.

Masters, C. J. (1977). Metabolic control and the microenvironment. *Current Topics in Cellular Regulation*, **12**, 75–105.

Rhodes, M. J. C. (1973). Co-factor specificity of plant alcohol dehydrogenases. *Phytochemistry*, **12**, 307–14.

Rhodes, M. J. C. & Wooltorton, L. S. C. (1968). A new fluorimetric method for the determination of pyridine nucleotides in plant material and its use in following changes in the pyridine nucleotides during the respiration climacteric in apples. *Phytochemistry*, **7**, 337–53.

Romell, L. G. (1924). *Jahrbuch für wissenschaftliche Botanik*, **65**, 739.

Wolstenholme, G. E. W. & O'Connor, C. M. (eds.) (1959). *Regulations of Cell Metabolism*. CIBA Foundation Symposium. London: Churchill.

I. ERICSON, P. GARDESTRÖM & A. BERGMAN

14 Isolation of leaf mitochondria and their role in photorespiration

The interest in leaf mitochondria has increased during the last decade, since it was discovered that, in the presence of atmospheric concentrations of oxygen and in light, most plants evolve carbon dioxide in a mitochondrial-dependent process called photorespiration (Zelitch, 1971). The intensified studies of leaf mitochondria will not only elucidate their role in photorespiration but also yield more detailed knowledge about metabolism in the photosynthesising cell.

Cell organelles are isolated with the aim of increasing our understanding of the situation in the cell *in vivo*. To achieve this, an organelle preparation must meet certain criteria with respect to purity and retained activities or functions. Different studies set different minimum requirements on the preparation. The main difficulty in purifying leaf mitochondria arises from the fact that they are a minor component of the photosynthesising cell. Chloroplasts constitute the largest proportion of the cell organelles in leaf tissue and heterogeneous chloroplast fragments produced upon homogenisation of the leaf are the main contaminant of the mitochondrial fraction after differential centrifugation. The first attempts to isolate leaf mitochondria failed to meet demands on purity and function, and gave little detailed information about leaf mitochondria. Then Douce, Moore & Neuburger (1977) designed a preparation medium that retained the respiratory functions of the mitochondria. Functional studies could thus be pursued on mitochondria prepared by differential centrifugation, although the preparation was heavily contaminated with chloroplast material. More recently several methods have been used to increase the purity of the mitochondria obtained by differential centrifugation. These include phase partition (Gardeström, Ericson & Larsson, 1978), Percoll density gradient centrifugation (Jackson *et al.*, 1979) and sucrose density gradient centrifugation (Arron, Spalding & Edwards, 1979). Also preparations from protoplasts may be useful (Nishimura, Douce & Akazawa, 1982). A very useful technique using three separation methods which separates the material according to different properties is outlined in Fig. 1. By this technique it is possible to obtain functional leaf mitochondria without contamination from chloroplast material (Bergman, Gardeström & Ericson, 1980).

Photorespiration is a complex process involving the participation of chloroplasts, peroxisomes and mitochondria. The generally accepted view, with emphasis on the role of mitochondria, is outlined in Fig. 2. It is thought that the first step in photorespiration is oxygenation of ribulose-1,5-bisphosphate (RuDP) instead of carboxylation by the enzyme RuDP carboxylase/oxygenase (Andrews & Lorimer, 1978). The physiological role of photorespiration is not known, but it can be seen as a process in which three-quarters of the carbon diverted from the Calvin cycle as glycolate is eventually recycled. Energetically photorespiration seems to be wasteful, since both carbon dioxide and ammonia evolved in the mitochondria need energy to be re-utilised. It may thus be a way for the plant to get rid of too much light energy (Andrews & Lorimer, 1978; Heber & Krause, 1980).

The ability to convert glycine to serine is missing in non-photosynthetic tissues such as root, stalk and leaf veins (Gardeström, Bergman & Ericson, 1980), and also in etiolated tissues. The evolution of the activity in etiolated tissue is light-dependent and seems to follow the increase in activity of the peroxisomal enzyme glycolate oxidase and not the increase in chlorophyll

Fig. 1. A schematic presentation of a combined preparation procedure, using different separation methods, which separates according to different properties of the membrane particles. For further details see Bergman *et al.* (1980).

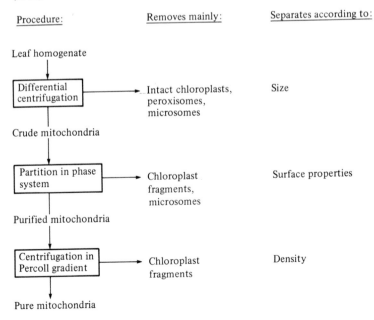

content (Arron & Edwards, 1980). C_4 plants have very low photorespiration measured as carbon dioxide evolution. In most cases, however, mitochondria from bundle sheath cells (where RuDP carboxylase is located) do convert glycine into serine, but the released carbon dioxide is efficiently refixed (Rathnam, 1979).

The mechanism for transport of glycine across the mitochondrial membrane is not clear, but glycine might penetrate in a ring form without the need for specific carriers (Day & Wiskich, 1980).

The conversion of glycine to serine involves two enzymes: glycine decarboxylase bound to the inner mitochondrial membrane (Moore *et al.*, 1977), and serine hydroxymethyltransferase, a matrix enzyme (Woo, 1979). The products from the glycine decarboxylase reaction are carbon dioxide, ammonia, methylene tetrahydrofolate and NADH (Bird *et al.*, 1972). The enzyme has not been purified from leaves, but the products are the same as for the better studied animal glycine decarboxylase (Motokawa & Kikuchi, 1974). Some important difference may, however, exist in the mechanism, as the leaf enzyme loses its activity when the membrane integrity is destroyed (Woo, 1979) while the animal enzyme does not. The NADH produced in the reaction can be oxidised by the mitochondrial electron transport chain. This oxidation is coupled to the production of three molecules of ATP. The NADH can alternatively reduce oxaloacetate to malate, a possible way of supplying NADH to the peroxisomes via a malate/oxaloacetate shuttle.

The glycine decarboxylation is believed to be the major reaction responsible for carbon dioxide evolution in photorespiration, though other reactions might also contribute. Glyoxalate can, for example, be peroxidated to carbon

Fig. 2. The main pathways of photorespiratory metabolism (the stoichiometry is not considered). GA, glycerate; GL, glycolate; GLY, glycine; GO, glyoxalate; HP, hydroxypyruvate; MAL, malate; OAA, oxaloacetate; PGA, 3-phosphoglycerate; PGL, phosphoglycolate; RuDP, ribulose-1,5-bisphosphate; SER, serine.

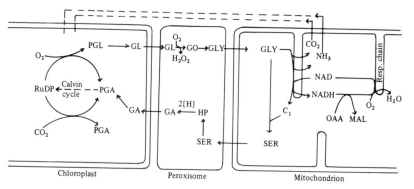

dioxide in the peroxisomes (Grodzinski & Butt, 1977). Refixation of photo-respiratory carbon dioxide occurs in the chloroplasts and makes it difficult to measure photorespiration as carbon dioxide release.

Ammonia produced in the glycine decarboxylation is not considered to be refixed in the mitochondria. The activity of glutamine synthetase is very low in mitochondria and the mitochondrial enzyme glutamate dehydrogenase which converts oxoglutarate to glutamate was too high a K_m for ammonia to be functional *in vivo*. The refixation can occur in the chloroplast by means of the glutamine synthetase (GS)/glutamine–oxoglutarate aminotransferase (GOGAT) system (Bergman, Gardeström & Ericson, 1981).

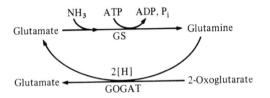

GS is also found in the cytoplasm but GOGAT is only found in the chloroplasts (Wallsgrove, Lea & Miflin, 1979). In spinach, however, GS is exclusively distributed to the chloroplasts (Hirel *et al.*, 1982). The importance of the GOGAT systems for the refixation of photorespiratory ammonia has been shown in studies with mutants deficient in leaf GOGAT activity (Somerville & Ogren, 1980).

References

Andrews, T. J. & Lorimer, G. H. (1978). Photorespiration – still unavoidable. *FEBS Letters*, **90**, 1–9.

Arron, G. P. & Edwards, G. E. (1980). Light-induced development of glycine oxidation by mitochondria from sunflower cotyledons. *Plant Science Letters*, **18**, 229–35.

Arron, G. P., Spalding, M. H. & Edwards, G. E. (1979). Isolation and oxidative properties of intact mitochondria from the leaves of *Sedum praealtum*. *Plant Physiology*, **64**, 182–6.

Bergman, A., Gardeström, P. & Ericson, I. (1980). Method to obtain a chlorophyll-free preparation of intact mitochondria from spinach leaves. *Plant Physiology*, **66**, 442–5.

Bergman, A., Gardeström, P. & Ericson, I. (1981). Release and refixation of ammonia during photorespiration. *Physiologia Plantarum*, **53**, 528–32.

Bird, I. F., Cornelius, M. J., Keys, A. J. & Whittingham, C. P. (1972). Oxidation and phosphorylation associated with the conversion of glycine to serine. *Phytochemistry*, **11**, 1587–98.

Day, D. A. & Wiskich, J. T. (1980). Glycine transport by pea leaf mitochondria. *FEBS Letters*, **112**, 191–4.

Douce, R., Moore, A. L. & Neuburger, M. (1977). Isolation and oxidative properties of intact mitochondria isolated from spinach leaves. *Plant Physiology*, **60**, 625–8.

Gardeström, P., Bergman, A. & Ericson, I. (1980). Oxidation of glycine via the respiratory chain in mitochondria prepared from different parts of spinach. *Plant Physiology*, **65**, 389–91.

Gardeström, P., Ericson, I. & Larsson, C. (1978). Preparation of mitochondria from green leaves of spinach by differential centrifugation and phase partition. *Plant Science Letters*, **13**, 231–9.

Grodzinski, B. & Butt, V. S. (1977). The effect of temperature on glycollate decarboxylation in leaf peroxisomes. *Planta, Berlin*, **133**, 261–6.

Heber, U. & Krause, G. H. (1980). What is the physiological role of photorespiration? *Trends in Biochemical Sciences*, **5**, 32–4.

Hirel, B., Perrot-Rechemann, C., Suzuki, A., Vidal, J. & Gadal, P. (1982). Glutamine synthetase in spinach leaves. *Plant Physiology*, **69**, 983–7.

Jackson, C., Dench, J. E., Hall, D. O. & Moore, A. L. (1979). Separation of mitochondria from contaminating subcellular structures utilizing silica sol gradient centrifugation. *Plant Physiology*, **64**, 150–3.

Moore, A. L., Jackson, C., Halliwell, B., Dench, J. E. & Hall, D. O. (1977). Intramitochondrial localization of glycine decarboxylase in spinach leaves. *Biochemical and Biophysical Research Communications*, **78**, 483–91.

Motokawa, Y. & Kikuchi, G. (1974). Glycine metabolism by rat liver mitochondria. *Archives of Biochemistry and Biophysics*, **164**, 634–40.

Nishimura, M., Douce, R. & Akazawa, T. (1982). Isolation and characterization of metabolically competent mitochondria from spinach leaf protoplasts. *Plant Physiology*, **69**, 916–20.

Rathnam, C. K. M. (1979). Metabolic regulation of carbon flux during C_4 photosynthesis. II. *In situ* evidence for refixation of photorespiratory CO_2 by C_4 phosphoenolpyruvate carboxylase. *Planta, Berlin*, **145**, 13–23.

Somerville, C. R. & Ogren, W. L. (1980). Inhibition of photosynthesis in *Arabidopsis* mutants lacking leaf glutamate synthetase activity. *Nature, London*, **286**, 257–9.

Wallsgrove, R. M., Lea, P. J. & Miflin, B. J. (1979). Distribution of the enzymes of nitrogen assimilation within the pea leaf cell. *Plant Physiology*, **63**, 232–6.

Woo, K. C. (1979). Properties and intramitochondrial localization of serine hydroxymethyltransferase in leaves of higher plants. *Plant Physiology*, **63**, 783–7.

Zelitch, I. (1971). *Photosynthesis, Photorespiration and Plant Productivity*. New York & London: Academic Press.

M. J. EARNSHAW & A. COOKE

15 The role of cations in the regulation of electron transport

The experimental work to be discussed in this article has, of necessity, been carried out with isolated mitochondria. The precise ionic composition of higher plant cell organelles and cytosol *in vivo* is still in doubt, but the isolation procedures employed undoubtedly alter intramitochondrial ion content. Consequently, extrapolation of work with isolated mitochondria to the cellular situation must be made with care.

Before considering the role of cations in the energy-linked functions of plant mitochondria, it is necessary to appreciate that, in common with other biological membranes, mitochondria exhibit two kinds of passive interactions with cations. Firstly, non-specific binding occurs to negatively charged membrane sites which, e.g. for calcium (Ca^{2+}) are of both lower ($K_d \simeq 600 \ \mu M$, where K_d is the dissociation of the metal–ligand complex $= [M^+][L^-]/[ML]$) and higher affinity ($K_d \simeq 50 \ \mu M$). The binding of Ca^{2+}, and other divalent cations, to the low-affinity sites produces a mitochondrial contraction possibly as a result of a membrane conformational change (Earnshaw, 1978). Secondly, non-bound cations can screen the fixed negative charges at the membrane surfaces by forming a diffuse layer and thus lower the surface potential (Johnston, Møller & Palmer, 1979). There is probably also a structural role for divalent cations in maintaining mitochondrial membrane integrity. For example, osmotic damage to the outer membrane of plant mitochondria is promoted by the cation chelator ethylenediamine tetra-acetic acid (EDTA) but reduced by magnesium ions (Hanson & Day, 1980).

Distinct roles of cations in regulating electron transport of plant mito-chondria occur in the oxidation of exogenous pyridine nucleotides and as a result of cation transport events.

Oxidation of exogenous pyridine nucleotides

Most plant mitochondria, unlike vertebrate mitochondria, are able to oxidise exogenous NADH by a pathway resistant to rotenone and piericidin A but sensitive to antimycin A and which by-passes the first coupling site. The oxidation of exogenous NADH is carried out with a flavoprotein, probably located on the outer surface of the inner membrane, which donates electrons to ubiquinone (Palmer, 1976). Exogenous NADH oxidation is non-competitively inhibited by chelators such as ethylene glycol bis(β-aminoethyl)-N,N^1-tetraacetic acid (EGTA) and EDTA providing the chelator is present before the initiation of NADH oxidation. Activity can be recovered by the addition of a range of divalent cations but recent work has shown that Ca^{2+} plays a specific role in exogenous NADH oxidation (Møller, Johnston & Palmer, 1981). It should be emphasised here that neither chelators nor divalent cations have a direct effect on the oxidation of tricarboxylic acid (TCA) cycle intermediates. However, if EGTA, for example, is added after the commencement of NADH oxidation then much less inhibition of oxidation occurs. Since EGTA does not penetrate the inner membrane of intact mitochondria, it is likely that the mitochondrial Ca^{2+} is initially external to the inner membrane permeability barrier and that the act of respiration moves the Ca^{2+} to a site inaccessible to EGTA. Energy-linked Ca^{2+} transport (see later) is probably not involved since Ca^{2+} movement occurs in the presence of an uncoupler (Møller *et al.*, 1981). A recent suggestion is that Ca^{2+} may induce a conformational change in the flavoprotein which locks Ca^{2+} into the active site where it is inaccessible to EGTA (Møller *et al.*, 1981).

There is also a separate, non-specific role for cations in promoting exogenous NADH oxidation for which two, not necessarily mutually exclusive, explanations have been put forward. In the case of potassium (K^+)-stimulated NADH oxidation, it has been proposed that K^+ induces the release of membrane-bound Ca^{2+} thus increasing Ca^{2+} activity at the flavoprotein (Hanson & Day, 1980). Secondly, stimulation of NADH oxidation by a range of inorganic and organic cations of differing valencies has been suggested to be due to charge screening. This would decrease the inner membrane surface potential allowing the local concentration of NADH, which is negatively charged at physiological pH, to increase at the external flavoprotein (Johnston *et al.*, 1979).

Plant mitochondria, again unlike vertebrate mitochondria, are also able to oxidise exogenous NADPH via the respiratory chain. There are several features in common with the exogenous NADH oxidation pathway such as insensitivity to rotenone and a by-pass of the first coupling site, but NADPH oxidation is more sensitive to chelators and sulphydryl reagents. The inhibition of NADPH oxidation caused by chelators can be relieved by Ca^{2+}, strontium (Sr^{2+}) or manganese (Mn^{2+}). The evidence, therefore, suggests the existence of a separate flavoprotein for NADPH oxidation which is probably also located on the outer surface of the inner membrane (Arron & Edwards, 1980).

Energy-linked cation transport

The outer mitochondrial membrane appears to be freely permeable to most solutes with a molecular weight of less than *c.* 13000 daltons. The inner membrane, however, is relatively impermeable to most ions but is readily permeable to water. Ion transport across the inner membrane can be driven by either respiration or ATP hydrolysis leading to increased energy expenditure and resulting in elevated respiration or ATPase activity. The primary energy-conserving event is now generally, although not universally, thought to be the electrogenic separation of hydrogen (H^+) and hydroxyl (OH^-) ions across the inner mitochondrial membrane as developed by Mitchell in the Chemiosmotic Hypothesis (Mitchell, 1979). The resultant electrochemical proton gradient or proton-motive force (Δp) can be defined as:

$$\Delta p = \Delta \psi - Z\Delta pH,$$

where $\Delta \psi$ is the electrical potential difference and $-Z\Delta pH$ is the chemical potential difference across the membrane ($Z = 2.303\ RT/F$). The proton-motive force can thus drive either ATP synthesis or ion transport as direct energetic alternatives.

Respiration-driven salt transport is inhibited by uncouplers, due to collapse of ΔpH, but is unaffected by inhibitors of the ATPase such as oligomycin. Influx pumping is generally considered to involve the electroneutral exchange of OH^- for an anion through an antiporter (Fig. 1), although this is experimentally indistinguishable from a H^+/anion$^-$ symporter. The operation of the antiporter, therefore, discharges ΔpH but maintains $\Delta \psi$ which drives cation transport via an electrophoretic uniport. Efflux pumping of cations, on the other hand, is suggested to occur on a cation$^+$/H^+ antiporter with the anion distributing in response to $\Delta \psi$ (Fig. 1).

A straightforward example of influx pumping in plant mitochondria is provided by the addition of an oxidisable substrate (NADH is generally preferred as it is not transported) to mitochondria in a medium containing potassium phosphate. Rapid salt influx results in osmotic swelling and a small stimulation of respiration. Phosphate (P_i) transport occurs on the P_i^-/OH^- antiporter which is specifically inhibited by low concentrations of the sulphydryl reagent mersalyl. However, little monovalent cation specificity occurs and there is doubt as to whether the 'uniport' is a non-specific cation carrier or simply represents the lipid domain of the inner membrane. The cation uniport is presumably rate-limiting since the addition of valinomycin (a K^+-binding ionophore) dramatically increases salt influx and osmotic swelling resulting in a large increase in respiration. In metabolic terms, the P_i^-/OH^- antiporter provides internal P_i as a substrate for ATP synthesis and can drive the net accumulation of TCA cycle intermediates via a series of exchange carriers (Hanson & Day, 1980). The cation uniport can, therefore, be viewed as providing internal charge compensation during the accumulation of metabolite anions. There is also evidence to suggest that steady-state osmotic swelling in potassium phosphate represents a balance between the influx pumping described above and efflux pumping (Fig. 1) involving a K^+/H^+ antiporter with P_i efflux occurring via a separate pathway to the P_i^-/OH^- antiporter (Hanson & Day, 1980).

Both plant and animal mitochondria will also transport Ca^{2+} and certain other divalent cations. The rapid respiration-driven influx of Ca^{2+} in vertebrate mitochondria is enhanced by P_i and involves a large stimulation of respiration and accompanying H^+ ejection. Both lanthanum (La^{3+}) and ruthenium red specifically inhibit Ca^{2+} influx, which is generally considered to be via an electrophoretic Ca^{2+} uniport (Bygrave, 1977). Ca^{2+} influx in plant mitochondria invariably occurs at a slower rate and with a lower affinity. The associated respiratory stimulation is variable in magnitude but stoichiometric

Fig. 1. Possible mechanisms for respiration-driven salt pumping in mitochondria where A is an antiporter driven by ΔpH and U is a uniport driven by $\Delta\psi$. M, inner membrane of the mitochondrion. (Based on Hanson & Day, 1980.)

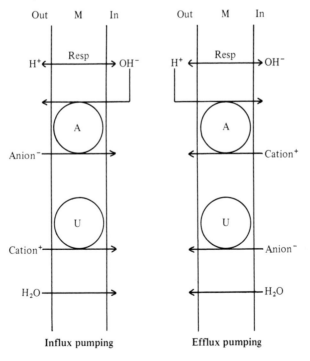

Influx pumping   Efflux pumping

H^+ ejection does occur. The effects of lanthanides and ruthenium red are inconsistent, leading to postulated mechanisms which vary from a slow electrophoretic Ca^{2+} influx (Russell & Wilson, 1978) to lack of a Ca^{2+} uniport in corn mitochondria where there is an obligatory dependence on accompanying P_i transport possibly via the P_i^-/OH^- antiporter (Hanson & Day, 1980). Calcium transport in the presence of P_i leads to calcium phosphate precipitation in the matrix space of both vertebrate and plant mitochondria.

As mentioned earlier, it is difficult to be precise about the role of cations in cellular respiration. However, it has been suggested that salt influx and efflux pumps (Fig. 1) may be responsible *in vivo* for controlling the volume of both plant and animal mitochondria (Hanson & Day, 1980). Isolated vertebrate mitochondria possess, in addition to the Ca^{2+} uniport discussed previously, a Ca^{2+} efflux pathway which in some mitochondrial types is a Na^+/Ca^{2+} antiporter (Nicholls & Crompton, 1980). Cyclic Ca^{2+} transport may thus provide a possible control system for the distribution of cellular Ca^{2+} which could lead to a number of metabolic consequences. For example, recent evidence suggests that the concentration of intramitochondrial Ca^{2+} may control the activity of some of the TCA cycle dehydrogenases (Nicholls & Crompton, 1980). Alterations in the cytosol Ca^{2+} concentration effected by the mitochondria may also regulate pyruvate kinase activity (Bygrave, 1977) and play a role in muscle contraction–relaxation in conjunction with the sarcoplasmic reticulum. The Ca^{2+}-sequestering role of mitochondria has also been invoked in avian shell and vertebrate bone formation.

There is scant information concerning Ca^{2+} efflux pathways in plant mitochondria which, together with the lack of a comparable Ca^{2+} uniport to vertebrate mitochondria, may reflect the different physiological requirements of plants. It is possible that plant mitochondria may act as Ca^{2+}-sequestering sites in cell wall formation. However, a more intriguing suggestion is that changes in Ca^{2+} concentration in the cytosol brought about by the mitochondria could directly affect exogenous NADH and NADPH oxidation (Palmer, 1976). In vertebrate cells, reducing equivalents from NADH produced by glycolysis are supplied to the mitochondria by means of complex anion shuttle systems. Direct regulation by Ca^{2+} of the cytosol [NAD(P)H]/[NAD(P)$^+$] ratio in the plant cell would have profound effects on the rates of glycolysis and the oxidative pentose phosphate pathway.

References

Arron, G. P. & Edwards, G. E. (1980). Oxidation of reduced nicotinamide adenine dinucleotide phosphate by potato mitochondria. *Plant Physiology*, **65**, 591–4.

Bygrave, F. L. (1977). Mitochondrial calcium transport. In *Current Topics in Bioenergetics*, vol. 6, ed. D. R. Sanadi, pp. 259–318. New York & London: Academic Press.

Earnshaw, M. J. (1978). The calcium-induced contraction of non-energised corn mitochondria. In *Plant Mitochondria*, ed. G. Ducet & C. Lance, pp. 159–66. Amsterdam: Elsevier.

Hanson, J. B. & Day, D. A. (1980). Plant mitochondria. In *The Biochemistry of Plants*, vol. 1, ed. N. E. Tolbert, pp. 315–58. New York & London: Academic Press.

Johnston, S. P., Møller, I. M. & Palmer, J. M. (1979). The stimulation of exogenous NADH oxidation in Jerusalem artichoke mitochondria by screening of charges on the membranes. *FEBS Letters*, **108**, 28–32.

Mitchell, P. (1979). Compartmentation and communication in living systems. Ligand conduction: a general catalytic principle in chemical, osmotic and chemiosmotic reaction systems. *European Journal of Biochemistry*, **95**, 1–20.

Møller, I. M., Johnston, S. P. & Palmer, J. M. (1981). A specific role for Ca^{2+} in the oxidation of exogenous NADH by Jerusalem artichoke (*Helianthus tuberosus*) mitochondria. *Biochemical Journal*, **194**, 487–95.

Nicholls, D. G. & Crompton, M. (1980). Mitochondrial calcium transport. *FEBS Letters*, **111**, 261–8.

Palmer, J. M. (1976). The organization and regulation of electron transport in plant mitochondria. *Annual Review of Plant Physiology*, **27**, 133–57.

Russell, M. J. & Wilson, S. B. (1978). Calcium transport in plant mitochondria. In *Plant Mitochondria*, ed. G. Ducet & C. Lance, pp. 175–82. Amsterdam: Elsevier.

A. PRADET & P. RAYMOND

16 Adenylate energy charge, an indicator of energy metabolism

Adenine nucleotide concentration ratios or adenylate energy charge?

Studies on enzyme regulation revealed the opposite effect of ATP on the one hand and ADP or AMP on the other, on the activity of some enzymes, and this suggested that regulation of enzyme activity could depend on [ATP]/[ADP] or [ATP]/[AMP] ratios rather than on the individual concentration of any of the nucleotides.

According to the model of enzyme regulation proposed by Atkinson (1968), regulation occurs by the competition of ATP and ADP or AMP for a catalytic or regulatory site on the enzyme and depends on the affinity of the site for each nucleotide and on the value of the concentration ratios.

For these reasons, enzyme activities or metabolic states could well be described as functions of concentration ratios of the different forms of the adenine nucleotides. However, using an analogy with a car battery, for which it is more relevant to know the charge than the potential, Atkinson (1968) expressed the energy state of the cell as the mole fraction of adenylate energy-rich compounds and called it 'adenylate energy charge', or AEC:

$$AEC = \frac{[ATP] + 0.5[ADP]}{[ATP] + [ADP] + [AMP]}.$$

This relation holds because in most living cells the enzyme adenylate kinase ($2\,ADP \rightleftharpoons ATP + AMP$) maintains the adenylate system at near-equilibrium (Bomsel & Pradet, 1968), so that β and γ pyrophosphate bonds of ATP are energetically equivalent.

Studies of enzyme activity *in vitro* showed that increasing AEC produced an activation of key enzymes from biosynthetic 'ATP-using' pathways. Conversely those from catabolic 'ATP-regenerating' pathways were inhibited. The responses were steeper in the range of the high AEC values, thus suggesting that the regulation of enzymes by AEC could explain stabilisation of the value around 0.85.

The use of this parameter was criticised for several reasons:

1. AEC is not directly correlated to free energy changes as is phosphate

potential: $[ATP]/[ADP] \times [P_i]$. However, phosphate concentration $[P_i]$ in a tissue is difficult to assess, particularly in vacuolised tissues, because of the presence of metabolically inactive phosphate pools which should be distinguished from the metabolically active pool.

2. In the range of high AEC values (from 0.8 to 1), where most *in vivo* values are found, small variations in AEC correspond to large variations in the ratios $[ATP]/[ADP]$ or $[ATP]/[AMP]$. The invariance of AEC found in most 'normally' metabolising tissues may then hide great changes in metabolic states.

3. The hypothesis of the regulation of enzymes by AEC has been discussed by Purich & Fromm (1973), who found that factors such as pH, concentration of magnesium or reaction products and the adenylate pool size, modify enzyme responses to AEC so strongly that data obtained *in vitro* are not sufficient to support the hypothesis that this parameter plays a regulatory role *in vivo*. Moreover, it is questionable whether the steepness of enzyme responses to high AEC is of any value in explaining regulation: in fact, the enzyme response curve merely runs parallel to the variation in the ratio $[ATP]/[ADP]$ versus AEC. Activity therefore varies almost linearly with the $[ATP]/[ADP]$ ratio and is not steeper in the range of $[ATP]/[ADP]$ values generally found *in vivo* ($[ATP]/[ADP]$ varies from 3.8 to 10.8 as AEC varies from 0.85 to 0.95, when the apparent equilibrium constant, K'_{app}, of the adenylate kinase reaction is one). The steep response argument thus cannot be used to explain the stabilisation of $[ATP]/[ADP]$ ratio or the related AEC in the range found *in vivo*. Alternatively, it may be concluded that the reported responses of enzymes to AEC would regulate the AEC as effectively at any value within the range studied (0.3–0.95).

Control of adenylate energy charge *in vivo*

Experimental errors arise from the fact that ATP turnover occurs in a matter of seconds. Three possible causes of experimental errors, and ways of avoiding them, are as follows:

1. Stopping enzyme activity is a crucial step in obtaining the exact value of adenine nucleotide concentrations at the chosen time in the experimental process. Quick freezing of the tissue permits this.

2. To inactivate plant phosphatase is difficult and Bieleski (1962) noticed that the treatment of tissues by an acid dissolved in an organic solvent produced the highest ATP values. In spite of this, methods of extraction using boiling water, boiling buffers or aqueous perchloric acid are still employed, yielding low AEC values.

3. The method most often used to estimate the adenine nucleotide concentration in plant extracts is the Strehler firefly assay adapted for ADP

and AMP estimation (Pradet, 1967). AEC can also be estimated by labelling the tissues with [^{32}P]phosphate or [^{14}C]adenine and then separating the phosphate esters chromatographically. When metabolically inactive pools of nucleotides are present (Hourmant, Pradet & Penot, 1979) this method gives a better estimation of the AEC because only the metabolic pools are labelled: stored or bound nucleotides are not labelled and do not interfere in the calculation.

In spite of the possibility that different compartments have different AEC values, estimation of AEC *in vivo* gives a good approximation of nucleotide ratios in the cytoplasm (Krebs, 1973).

It was asserted more than a decade ago (Bomsel & Pradet, 1968; Pradet, 1969) that in actively metabolising tissues AEC is usually high. This observation was then extended to microorganisms (Chapman, Fall & Atkinson, 1971). In plants, however, low values have often been recorded and the idea has sometimes been put forward that AEC in plants is not regulated in the same way as in other living organisms. However, it has now been established by using improved techniques that AEC values of aerated and non-starved plant tissues are high.

The ability to control experimentally the AEC *in vivo* would permit us to understand the metabolic significance of the parameter. A correlation between metabolic activity and AEC value has sometimes been postulated. However, Chapman & Atkinson (1977) consider that if such a correlation existed, it would be in such a narrow range of AEC values that it could not be demonstrated experimentally.

ATP production and utilisation are so closely connected that a control of metabolic activity can be achieved by methods which limit any one or both of these functions. Sugar starvation (Saglio & Pradet, 1980) or a decrease in temperature (Pradet, 1969) under normoxic conditions can induce a marked limitation of respiration without variation in the AEC (Fig. 1*a*). Decreasing anabolic processes by an inhibition of protein synthesis with cycloheximide (Coccuci & Marre, 1973) or through ammonium starvation slightly increased the AEC, while stimulating protein synthesis decreased it. These changes are small but as they are in the upper part of the scale they correspond to great changes in metabolic states (Fig. 1*b*).

Large decreases in AEC were first observed in germinating lettuce seeds under oxygen partial pressures that limited their respiration. The AEC decreased to intermediate values between that in air (0.9) and that under anoxia (0.3). These values remained stable for hours and were correlated with the residual respiratory activity (Raymond & Pradet, 1980) (Fig. 1*c*). It is well known that when cells are submitted to anoxia their AEC drops to values ranging from 0.8 to 0.2. When the glycolytic rate was controlled by the concentration of sugars, a straight correlation between glycolytic rate and AEC values was established in blood platelets in the presence of cyanide (Akkerman & Gorter, 1980) and in maize root tips under anoxia (Saglio, Raymond & Pradet, 1980). Low fermentation activity under anoxia corresponds to low AEC values while high fermentation rates maintain high AEC values (Fig. 1c).

Consequently AEC can be used to estimate metabolic fluxes during a limitation of the rate of ATP regeneration. Moreover, it appears that AEC is regulated in higher plants in the same way as in animal tissues.

Is energy charge an important regulatory parameter?

According to Atkinson (1968), regulation by adenine nucleotides is the simplest and most efficient system for adjusting the rates of energy-producing and energy-consuming pathways to each other. In fact, adenine nucleotides are common to all metabolic pathways and so they are potentially

Fig. 1. Generalisation of the effects on AEC value of a reduction of energy metabolism by factors which affect simultaneously ATP-utilising (U) and ATP-regenerating (R) pathways (*a*), or primarily U pathways (*b*), or primarily R pathways (*c*).

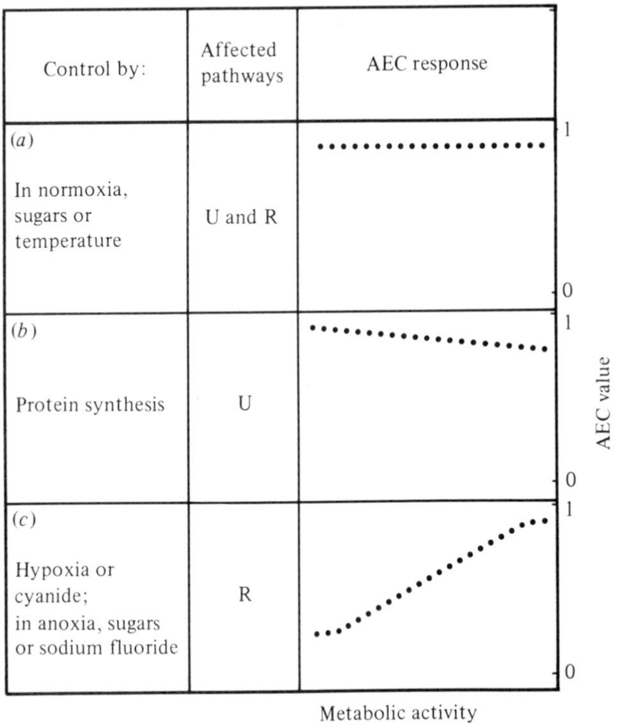

Control by:	Affected pathways	AEC response
(*a*) In normoxia, sugars or temperature	U and R	
(*b*) Protein synthesis	U	
(*c*) Hypoxia or cyanide; in anoxia, sugars or sodium fluoride	R	

Metabolic activity

good regulators of the metabolic network. The apparent invariance of AEC strengthened the view that the adenylate ratios were key factors in metabolic regulation. However, the recent demonstration that AEC can be stabilised not only at high values but also at intermediate and low values within its range (Raymond & Pradet, 1980) shows that no special status in cellular homeostasis should be assigned to the adenylate system.

Pyridine nucleotides also have a central role in metabolism – as redox transducers – and are part of most metabolic pathways (Atkinson, 1968). They could then participate in the regulation as easily as adenylates.

If, as proposed by Krebs (1973), equilibria form the basic framework of metabolism, then any variation in the ratios of adenine or pyridine nucleotide is coupled to variations in concentration of many other metabolites. The flux in any given pathway could be varied quickly either through stoichiometric relation (Sel'kov, 1975) or through the effect on a regulatory site of any metabolite involved in an equilibrium with adenine nucleotides. This could explain how AEC is regulated in the same way in higher plants as it is in other organisms in spite of the fact that phosphofructokinase from higher plants is not sensitive to adenylate ratios (Turner & Turner, 1975).

Thus, adenine nucleotide ratios are undoubtedly a part of a regulatory system, though not necessarily the most important one. However, variations in their ratios prove useful in characterising some cellular metabolic states. The experimental control of their values *in vivo* would permit studies of their regulatory role. These problems are studied in more detail elsewhere (Pradet & Raymond, 1983).

References

Akkerman, J. W. N. & Gorter, G. (1980). Relation between energy production and adenine nucleotide metabolism in human blood platelets. *Biochimica et Biophysica Acta*, **590**, 107–16.

Atkinson, D. E. (1968). The energy charge of the adenylate pool as a regulatory parameter. Interaction with feedback modifiers. *Biochemistry*, **7**, 4030–4.

Bieleski, R. L. (1962). The problem of halting enzyme action when extracting plant tissues. *Analytical Biochemistry*, **9**, 431–42.

Bomsel, J. L. & Pradet, A. (1968). Study of adenosine 5'-mono, di and triphosphates in plant tissues. IV. Regulation of the level of nucleotides, *in vivo*, by adenylate kinase: theoretical and experimental study. *Biochimica et Biophysica Acta*, **162**, 230–42.

Chapman, A. G. & Atkinson, D. E. (1977). Adenine nucleotide concentrations and turnover rates. Their correlation with biological activity in bacteria and yeast. In *Advances in Microbial Physiology*, vol. 15, ed. A. M. Rose & D. W. Tempest, pp. 254–307. New York & London: Academic Press.

Chapman, A. G., Fall, L. & Atkinson, D. E. (1971). Adenylate energy charge in *Escherichia coli* during growth and starvation. *Journal of Bacteriology*, **108**, 1072–86.

Coccuci, M. C. & Marre, E. (1973). The effect of cycloheximide on respiration, protein synthesis and adenosine nucleotide levels in *Rhodotorula gracilis*. *Plant Science Letters*, **1**, 293–301.

Hourmant, A., Pradet, A. & Penot, M. (1979). Action de la benzylaminopurine sur l'absorption du phosphate et le métabolisme des composés phosphorylés des disques de tubercule de pomme de terre en survie. *Physiologie Végétale*, **17**, 483–99.

Krebs, H. A. (1973). Pyridine nucleotides and rate control. *Symposia of the Society for Experimental Biology*, **27**, 299–318.

Pradet, A. (1967). Etude des adenosine 5′-mono, di et tri-phosphates dans les tissus végétaux. I. Dosage enzymatique. *Physiologie Végétale*, **5**, 209–21.

Pradet, A. (1969). Etude des adenosine 5′-mono, di et tri-phosphates dans les tissus végétaux. V. Effet *in vivo* sur le niveau de la charge énergétique d'un déséquilibre induit entre fourniture et utilisation de l'énergie dans les semences de laitue. *Physiologie Végétale*, **7**, 261–5.

Pradet, A. & Raymond, P. (1983). Adenine nucleotide ratios and adenylate energy charge in energy metabolism. *Annual Review of Plant Physiology*, **34**, 199–224.

Purich, D. L. & Fromm, H. J. (1973). Additional factors influencing enzyme responses to the adenylate energy charge. *Journal of Biological Chemistry*, **248**, 461–6.

Raymond, P. & Pradet, A. (1980). Stabilization of adenine nucleotide ratios at various values by an oxygen limitation of respiration in germinating lettuce (*Lactuca sativa*) seeds. *Biochemical Journal*, **190**, 39–44.

Saglio, P. H. & Pradet, A. (1980). Soluble sugars, respiration and energy charge during aging of excised maize root tips. *Plant Physiology*, **66**, 516–19.

Saglio, P. H., Raymond, P. & Pradet, A. (1980). Metabolic activity and energy charge of excised maize root tips under anoxia: control by soluble sugars. *Plant Physiology*, **66**, 1053–7.

Sel'kov, E. E. (1975). Stabilization of energy charge, generation of oscillation and multiple steady states in energy metabolism as a result of purely stoichiometric regulations. *European Journal of Biochemistry*, **59**, 151–7.

Turner, J. F. & Turner, D. H. (1975). The regulation of carbohydrate metabolism. *Annual Review of Plant Physiology*, **26**, 159–86.

INDEX